T0249856

ENGINEERING MECHANICS OF POLYMERIC MATERIALS

Theories, Properties, and Applications

ENGINEERING MECHANICS OF POLYMERIC MATERIALS

Theories, Properties, and Applications

Gabil Garibxan ogli Aliyev, DSc
and Faig Bakhman ogli Naghiyev, DSc

Apple Academic Press

TORONTO NEW JERSEY

Apple Academic Press Inc.
3333 Mistwell Crescent
Oakville, ON L6L 0A2
Canada

Apple Academic Press Inc.
9 Spinnaker Way
Waretown, NJ 08758
USA

First issued in paperback 2021

Exclusive worldwide distribution by CRC Press, a member of Taylor & Francis Group

ISBN 13: 978-1-77463-284-0 (pbk)
ISBN 13: 978-1-926895-55-0 (hbk)

Library of Congress Control Number: 2013945478

Library and Archives Canada Cataloguing in Publication

Aliyev, G. G., author
Engineering mechanics of polymeric materials: theories, properties, and applications/by Dr. G.G. Aliyev and Prof. Faig Nagiyev.

Includes bibliographical references and index.
ISBN 978-1-926895-55-0
1. Polymeric composites--Mechanical properties. 2. Polymers--Mechanical properties.
I. Nağıyev, Faiq, author II. Title.

TA418.9.C6A45 2013 620.1'92 C2013-904993-2

Apple Academic Press also publishes its books in a variety of electronic formats. Some content that appears in print may not be available in electronic format. For information about Apple Academic Press products, visit our website at **www.appleacademicpress.com** and the CRC Press website at **www.crcpress.com**

ABOUT THE AUTHORS

Gabil Garibxan ogli Aliyev, DSc

Gabil Garibxan ogli Aliyev, DSc, is currently affiliated with the Institute of Mathematics and Mechanics at the Azerbaijan National Academy of Sciences, Baku, Azerbaijan. He has published several books in the field as well as over 150 journal articles and international patents.

Faig Bakhman ogli Naghiyev, DSc

Faig Bakhman ogli Naghiyev, DSc, is currently affiliated with the Azerbaijan State Marine Academy in Baku, Azerbaijan. He is also the Leading Research Officer, Ministry of Communications and Information Technologies of Azerbaijan Republic, Baku, Azerbaijan. He was formerly Professor of the Department of Development and Production of Oil Fields at Azerbaijan State Oil Academy, Baku, Azerbaijan. He is the author of over 100 articles and five books and more than 200 analytical reports. He has worked on many projects on fundamental and applied studies in hydro-mechanics and thermal physics of multiphase media, mathematics, computer programming, research studies in oil and gas sphere, power-engineering and technological complexes, education and sciences.

CONTENTS

LIST OF ABBREVIATIONS

PA	Polyamid-610
PEHD	Polyethylene of high density
PELD	Polyethylene of lower density
PP	Polypropylene
PS	Polystyrene shock-resistance

PREFACE

Large production of polymer materials in the world redoubled attention to the problem on their use in engineering in place of load-bearing structural elements. However, in some applied problems of engineering by using polymer materials, for example, their work in contact with a different kind of corrosive liquid and gassy medium, operational characteristics of industrial constructions made of them sharply deteriorate, and this leads to premature loss of their bearing capacity. This says that the classical methods of mechanical strength stability and dynamics analysis of constructions become unreliable. It should be noted that today, in designing engineering of constructions working in contact with corrosive liquid and gassy medium, the reliable engineering methods for strength analysis of constructions with regard to physico-chemical properties of materials are not available. This is connected with the fact that account of influence of diffusion processes of corrosive media in polymers on strength characteristics of constructions, especially of composite materials, introduces considerable difficulties. In this context, the following principle questions remain open: how do the physico-mechanical properties of polymer and composite materials and also the problem on strength of structural elements made of them depend on physico-chemical changes of the material? In particular, how is the quality and quantity character of dependence of serviceability of different constructions made of polymer materials and also their mechanical properties on the character of their contact with corrosive liquid and gassy media? The joint solution of diffusion equations of corrosive and gassy media to polymers, and the equations of mechanics of deformable body made of polymer and composite material, is a very complicated mathematical problem of connected type. At the same time, such an approach requires to know exact values of initial properties of the material, diffusion factor and boundary conditions of the problem. Otherwise, the physico-mathematical problem may become ill-posed.

In this monograph, theory of strength of laminated and reinforced structures made of polymer materials with regard to changeability of their physico-chemical properties is elaborated upon. To achieve this goal, an experimental-theoretical method on definition of physico-mechanical properties of polymers, composite materials and also polymerized bundles made of fibers with regard to change of physico-chemical properties of the material is suggested. On this basis, Hooke's generalized law with regard to changeability of physico-chemical properties of polymer material is suggested; a criterion of adhesive strength of laminated and sandwich reinforced pipes with regard to physico-chemical changeability of the structure's layers is suggested; stability criterion of one-dimensional structural elements with regard to swelling effect of materials is suggested; new mechanical effects such as changeability of mass force of a polymer material with regard to material's swelling effect buckling of columns under the action of only material's swelling effect were revealed. A mechanical effect, dependence of joint deformability of several polymer materials on the character of layer's swelling was experimentally and theoretically established.

We hope this monograph will be useful for research associates engaged in the problems of designing of construction from polymer and composite materials that are intended to work in corrosive liquid and gassy media. We think that the obtained scientific results may find its reflection in some physico-mathematical problems of biomechanics.

— Gabil Garibxan ogli Aliyev, DSc
and Faig Bakhman ogli Naghiyev, DSc

CHAPTER 1

INTERACTION OF PHYSICO–MECHANICAL AND PHYSICO–CHEMICAL PROPERTIES OF POLYMER MATERIALS

G. G. ALIYEV and F. B. NAGIYEV

CONTENTS

1.1 OBJECTIVE LAWS OF MASS-TRANSFER OF CORROSIVE LIQUID AND GASSY MEDIA IN POLYMER AND COMPOSITE MATERIALS AND SOLUTION METHODS OF DIFFERENTIAL TRANSFER EQUATIONS

The media transfer processes are based on diffusion phenomena, that is, spontaneous travel of atoms and molecules in consequence of their thermal motion. Depending on proceeding conditions of this process, there is an interdiffusion observed in the presence of concentration gradient or in the general case, the chemical potential gradient; secondly, a self-diffusion is observed when the above-mentioned ones are not available. Under interdiffusion, the diffusing particles flux is directed to the side of concentration decrease. As a result, the substances are distributed in the space, and local differences of potentials and concentrations are leveled. The characteristic quantity of such a process is the interdiffusion coefficient *D*. In the case of one-dimensional diffusion, the relation between the diffusing substance flux f_i and the concentration gradient $\partial \tilde{N}_i / \partial x$ in an isotropic medium at rest is described by Fick's differential equation:

$$(J_i)_R = -(D)_p \frac{\partial C}{\partial x} \tag{1.1}$$

According to this relation, *D* numerically equals the flux density with respect to the section *R* under the given concentration gradient. Simultaneously, *D* may be considered as velocity by which the system is capable to level the unit difference of concentrations.

In this way, we can define the problems that compose the mass-transfer content in the section of physics of polymers.

Firstly, to work out a method of experimental investigation of diffusion process, that is, to study kinetic objective laws of diffusion parameters change: either concentration distribution or the sizes of diffusion zone, or the amount of diffusing substance.

Secondly, to create a method for defining the diffusion coefficients on experimental data and the solution of the inverse problem-prediction of the process course if *D* and kinetic conditions are known. This group of questions is the subject of phenomenological or mathematical theory of diffusion. If at experimental measurements the physical methods are an instrument in the hands of researchers, then in phenomenological theory the apparatus of mathematical physics is such an instrument. Analytic equations connecting changes of these or other external parameters of registered in the test, of coordinate diffusion, with time, diffusion coefficient, and sample's sizes are obtained with the help of this apparatus.

Since, as a rule, the diffusion factor in these equations is an unknown quantity, then processing of experimental data with their help allows getting quantity information on its value, to establish correctness of the accepted mathematical model to the real process.

Thirdly, analysis of elementary act of translational travel of molecules or macromolecules on the basis of diffusion mechanism and theoretical calculation of partial coefficient of self diffusion of diffusing particle in a polymer medium. This is a problem of molecular-kinetic theory of diffusion.

Fourthly, explanation of some "strange" phenomena known as "negative diffusion" and "ascending diffusion", determination of direction of diffusion fluxes, calculation of thermodynamical parameters of the system by diffusion data.

All above-stated problems and directions are closely connected between themselves forming a united chain of investigations of diffusion processes.

Two fundamental laws of nature: the law of preservation of mass and the law of preservation and transformation of energy are on the basis of above-mentioned section of transfer phenomenon. Directedness of transfer processes is determined by the second law of classic thermodynamics the law of increase of entropy. The most common equations of mass and heat transfer are obtained on the basis of the mentioned laws.

In thermodynamics of irreversible processes, it was established (Grot and Mazur, 1967; Lykov, 1978; Sedov, 1970), that the product of the entropy increases velocity $\frac{dS}{dt}$ by absolute temperature T, and equals the sum of products of flux densities J_i by the appropriate thermodynamical moving forces X_i:

$$T\frac{dS}{dt} = \sum J_i X_i \tag{1.2}$$

According to this definition, thermodynamical forces of energy (heat) transfer X_T and the masses of the substance X_μ equal:

$$X_T = -\frac{1}{T}\nabla T, \quad X_\mu = -T\nabla\left(\frac{\mu}{T}\right) \tag{1.3}$$

Here, μ is thermodynamical potential of components.

By the thermodynamics of irreversible processes, the mass-transfer of the substance and also transfer of heat energy, electric charges and other factors of energy carriers that increase the system's entropy, in general case are determined by the action of all *n* thermodynamical forces:

$$J_i = \sum_{k=1}^{n} L_{ik} X_k \quad (i=1, 2, ..., n) \tag{1.4}$$

Thus, the mass-transfer of the substance, heat fluxes, electric charges transfer are interconnected and interdependent. Relation (1.4) valid for negligible deviations from the equilibrium state, is known as a system of Onzager linear equations and is the basis of thermodynamics of irreversible processes. The quantities L_{ik} called kinetic coefficients are connected with the relations $L_{ik} = L_{ki}$. As it follows from this relation, the mass transfer of the substance may be described by the system of interconnected equations. Quantity distribution of mass transfer includes the action not only of direct thermodynamical force (so-called diffusion concentration-of diffusion at the expense of concentration gradient of the substance), but also of all other forces; of the force that is proportional to temperature gradient (Soreh's effect); of electric field action force (electrophoresis effect); of gravitational, magnetic force and other fields action forces.

Non availability of these or other influences essentially simplifies system (1.4). So, for non isothermic conditions, where there are no other thermodynamic forces, besides the mentioned temperature gradients and substance concentration, we get the system of two equations:

$$J_1 = L_{11}X_1 + L_{12}X_2 = -\frac{L_{11}}{T}\nabla T - L_{12}T\left(\frac{\mu}{T}\right)$$

$$J_2 = L_{21}X_1 + L_{22}X_2 = -\frac{L_{21}}{T}\nabla T - L_{22}T\left(\frac{\mu}{T}\right) \tag{1.5}$$

where, J_1 and J_2 are the flux densities of heat energy and substance mass, respectively.

According to reciprocity principle, the cross coefficients in these equations that are valid both for stationary and non stationary conditions, are equal that in some cases simplifies the calculations. When in addition an electric field influences on the system, we get the system of three equations and so on.

On the other hand, in the absence of other fields except temperature field, on equation (1.5) we get the expression of the Fourier law for the heat flux:

$$J_1 = -\lambda \cdot gradT \tag{1.6}$$

where, $\lambda = {L_{11}}/{T}$ is a heat conductivity coefficient.

At non availability of the temperature gradient, from (1.5) we get the so called the first law of Fick:

$$J_2 = -D \cdot gradC \tag{1.7}$$

where, the diffusion coefficient $D = L_{22}T$ is a proportionality coefficient, and $C = \mu T$ is a substance concentration under the given temperature.

In the presence of some components, we write equation (1.7) for each of them. Thus, if the process of diffusion of binary medium or multicomponent liquid to solid phase is realized, then we should write an original equation for each component and the number of diffusion equations and their parameters becomes equal to the number of components. Only in some special cases, the diffusion process of some solution of liquids with close properties may be described by a mass-transfer equation containing some averaged parameters. But such a change needs special ground and experimental confirmation.

The experience shows that (Reytlinger, 1970; Sachs and Lubahn, 1946; Stepanov and Shlenscy, 1981) each of independent diffusion fluxes of multicomponent system are a homogeneous linear function of all independent concentration gradients except extremely high gradient. Consequently, for N types of particles (under these particles we can imply atoms of various liquids moving in fixed medium) we have:

$$J_i = -\sum D_{ik} gradC_k \quad (i = 2,3,\ldots,N) \tag{1.8}$$

Reading of indices is taken from 2, as it is supposed that the coordinate system is connected with a fixed continuum. According to the basic principles of thermodynamics of irreversible processes, the diffusion coefficients are independent of concentration gradient, but they are the functions of local parameters of the system's state: temperature, pressure, and concentration. Mechanical stresses, strains, and also their relations have an influence on medium diffusion coefficients in real materials including polymer materials.

Considering this influence equation (1.8) is the generalization of Fick's law for several components. In this form, it is usually written for stationary state on the basis of this equation, and also the law of preservation of mass, for non stationary state of each of the components in per unit volume, we get a differential mass-transfer equation:

$$\frac{\partial C_i}{\partial t} = div(\sum_{k=2}^{N} D_{ik} gradC_k) - div(C_i W), \quad (i = 2,3,\ldots,N) \tag{1.9}$$

In the last equation, there is an addend taking into account velocity W of a convective, or as it is called, molar mass-transfer. Unlike the diffusive mass-transfer, the convective mass-transfer in solid substance may be caused by pressure differential (by the Darcy law or by a stronger law) by centrifugal forces, capillary effect, and so on. The molar convective transfer process usually occurs in heterogeneous bodies and represents itself as medium's motion on micro- and macro cracks, cavities and other deficiencies of the structure. In conformity to plastic masses, such processes are often observed of glass reinforced plastics, asbestos-filled plastic type materials that differ by inhomogenity of the structure, but it may occur in unfilled materials with structural deficiency.

Equation (1.9) remains valid in the absence of chemical reactions. Under diffusion of liquid media into polymer materials, this equation is applicable for chemically inactive liquids not interacting with a polymer.

However, formally it is not difficult to take into account chemical interaction for media corrosive with respect to the given polymer. Such an interaction may be reduced to calculation of the *i*th substances in the equations of sources and flows. Then, it is necessary to complement equation (1.9) by the appropriate addends $Q_{\mu i}$:

$$\frac{\partial C_i}{\partial t} = div(\sum_{k=2}^{N} D_{ik}\, gradC_k) - div(C_i W) \pm Q_{\mu i} \qquad (1.10)$$

Here, the signs denote addition or decrease of the component because of chemical reactions. Obviously, the sum of volume densities of fluxes of all sources, and flows of the substance mass in the elementary volume equals zero.

$$\sum_{1}^{N} Q_{\mu i} = 0 \qquad (1.11)$$

Such an analysis of the system of mass- and heat-transfer equations both in non availability and in proceeding of chemical reactions in the fields of conservative forces are given in the monographs (Grot and Mazur, 1967; Kargin and Slonimsky, 1960; Lykov, 1978; Manin and Gromov, 1980; Malmeyster et. al., 1980; Ogibalov and Suvorova, 1965; Ogibalov et. al., 1972; Ogibalov et al., 1972; Reytlinger, 1970; Sachs and Lubahn, 1946). By analogy with the mass-transfer equation, the energy conservation equation (heat conductivity equation) has a similar structure.

$$c\rho \frac{\partial T}{\partial t} = div(\lambda \cdot gradT) - div(C_i W_i) \pm Q_i \qquad (1.12)$$

c is a mass heat capacity; λ is a heat conductivity factor; Q_i is a volume density of the heat flux of sources and flows.

In the practice of engineering calculations, it turns out that the cross effects (Sore's effect and diffusive heat conductivity) negligibly influence on mass-transfer. Therefore, in analytic solutions of problems, usually they are not taken into account and equations (1.10) and (1.12) become disconnected. While calculating the liquid concentration fields by numerical methods, realized by means of computers, account of the enlisted effects and also non linearity of the equations because of dependence of coefficients on variable quantities $x,y,z,t,T,$ does not introduce principal difficulties.

As a rule, the greatest difficulty is non availability of necessary experimental data, the number coefficients of equations determined with sufficient degree of reliability.

1.1.1 BOUNDARY CONDITIONS IN MASS AND HEAT TRANSFER PROBLEMS

For solving differential equations of mass and heat transfer, it is necessary to know the so called boundary conditions that include initial and boundary conditions of the specific problem given. At initial conditions, temperature and concentration distributions at the moment equal to zero are given in the form of coordinate function.

These conditions are written in the form:

$$T(x,y,z) = f_T(x,y,z)\Big|_{t=0},\quad C(x,y,z) = f_C(x,y,z)\Big|_{t=0} \tag{1.13}$$

where, the index n denotes the body's surface. In the specific but much extended case, the body's surface temperature may be constant in time and equal medium's temperature:

$$T_n = T_C \tag{1.14}$$

Such a condition may be attained for example, by means of automatic control of surface temperature. Then the first kind boundary condition will be of the form:

$$T_n = const \tag{1.15}$$

As it follows from the cited further solutions of transfer problems in finite dimensional bodies, for $T_n = const$ and $C_n = const$ as a result of gradual leveling of temperature and concentration, the substances tend to accept equilibrium values $T_n = T_\infty$, $C_C = C_\infty$ on the volume of the body. Unlike the temperature, the quantity C_∞ is most available for experimental definition. In this connection, further, instead of Cn we will take its equal value C_∞. Thus, in the case when a polymer material contacts with a liquid media, when it is stirred properly, *the first kind boundary conditions* are written in the form:

$$C_{\infty} = const \tag{1.16}$$

In the boundary conditions of second kind, the densities of heat q_T or mass q_M fluxes are given at each point of the body's surface as a time function:

$$q_T(t) = f_T(t)\ ,\ q_M(t) = f_M(t) \tag{1.17}$$

In the simplest case, the heat and mass flux densities are constant:

$$q_T = const\ ,\ q_M = const \tag{1.18}$$

The second kind boundary conditions are realized under heat transfer by the Stephan–Boltzmann law, and for mass exchange, for example by drying moist bodies, where at the first period of drying, the drying intensity is constant.

In the boundary conditions of the third kind, the temperature of environment (for example, for convective heat exchange) is given. According to Newton's law for convective heat exchange the heat flux through the surface is proportional to difference of temperatures of body surface T_n and of environment T_c:

$$-\lambda\left(\frac{\partial T}{\partial n}\right)_n = q_T(t) = \alpha[T_n(t) - T_c(t)] \tag{1.19}$$

where, n is normal to the surface. In the similar way we write the boundary conditions for mass exchange:

$$D\left(\frac{\partial C}{\partial n}\right)_n + \alpha_m[C_n(t) - C_c(t)] = 0 \tag{1.20}$$

where, α_m is a mass exchange coefficient.

The third order boundary conditions are realized under weak motion of external medium, that is, under negligibly intensive stirring. Under intensification of stirring, the relation $(\alpha/\lambda) \to \infty$ since $\alpha \to \infty$ and in the special case, from the third kind boundary conditions we get first order boundary conditions $C_{\infty} = const$.

The fourth kind boundary conditions correspond to ideal contact of the surface of two bodies. In this case:

$$T_{n1} = T_{n2},\ C_{n1} = C_{n2} \tag{1.21}$$

Besides, the equality on the contact surface should be provided by the equality of heat and diffusion flows:

$$\lambda_1 \left(\frac{\partial T}{\partial n}\right)_1 = \lambda_2 \left(\frac{\partial T}{\partial n}\right)_2,\ \left(\frac{\partial C}{\partial n}\right)_1 = D_2 \left(\frac{\partial C}{\partial n}\right)_2 \tag{1.22}$$

Periodic and antiperiodic influence of external medium with change of its concentration or properties is the special case of the mentioned conditions. Combinations of the indicated boundary conditions, for example, mixed conditions of second and third order, are also possible.

1.1.2 SOLUTION METHODS OF DIFFERENTIAL TRANSFER EQUATIONS

The system of differential equations of heat and mass transfer together with boundary conditions allow not only to get the most complete information on the principles of diffusing substance distribution in the material in the course of time but also to study influence of basic factors on such a distribution, such as sizes and configuration of body, intensity of external mass exchange, properties of diffusate by adsorbent media, and so on. In this case, empiric investigations or empiric character approximate methods cannot replace analytic methods based on fundamental laws of physics. Nevertheless, insufficient study of the processes of mass transfer of media in polymer materials requires thorough verification of solutions of appropriate problems. It should also be noted that the solutions of composite problems valid for one pair of polymer liquid may turn to be inapplicable to other pairs. This means that in this or other case, all the factors influencing the process and the problem solutions have not been précised and in this case they require appropriate precision. Consider the basic solution methods for transfer differential equations used at present (Stepanov and Shlenscy, 1981).

Exact analytic solutions of transfer differential equations, as a rule, are obtained by the method of separation of variables, methods of instantaneous sources, and also by means of the Green function. The operational methods on the Laplace transform are also widely used (Stepanov and Shlenscy, 1981). Cite basic results of the solutions of some transfer problems that represent the greatest interest for executing practical calculations accept the following denotation: C_0 is the concentration of diffusate in the body at initial time; $C_\infty = C_n$ is the equilibrium value of concentration; $C = \nabla M / M_0$ is the current value of medium's concentration that equals the ratio of increments of the mass at the given point (of volume) to its initial mass M_0. In the accepted denotations, Fick's diffusion equation for one-dimensional process has the form:

$$\frac{\partial C}{\partial t} = D\frac{\partial^2 C}{\partial x^2} \tag{1.23}$$

Cite the solution for first kind boundary conditions.

A semi-bounded body. The initial concentration of the medium $C_0 = 0$. At moment $t = 0$, the body's surface contacts with the liquid medium and the medium's concentration $C_n = C_\infty$ is established on the surface by a jump. Then, we describe the medium concentration distribution on the body by the law:

$$\frac{C_\infty - C(x,t)}{C_\infty - C_0} = erf\left(\frac{x}{2\sqrt{Dt}}\right), \tag{1.24}$$

Where, $erf\left(\frac{x}{2\sqrt{Dt}}\right)$ is the Gauss error function equal

$$?rf(u) = \frac{2}{\sqrt{u}}\int_0^u e^{-u^2} du \tag{1.25}$$

The quantity (mass) of the substance diffused into the body through the surface unit after a lapse of exposition time t equals:

$$\Delta M = 2\sqrt{\frac{Dt}{\pi}}(C_\infty - C_0)\cdot M_0 \quad (1.26)$$

An unbounded plate under symmetric influence of medium. The initial concentration of medium $C_0 = 0$. Medium concentration distribution on the plate's thickness at time *t* after beginning the exposition is subjected to the equation:

$$\theta = \frac{C_\infty - C(x,t)}{C_\infty - C_0} = \sum_{n=1}^{\infty}\frac{2}{\mu_n}(-1)^{n+1}\cos\mu_n\frac{x}{R}\cdot\exp(-\mu_n^2 F_0) \quad (1.27)$$

where, *R* is the half of plate's thickness; $F_0 = \frac{Dt}{R^2}$ is a dimensionless complex (Fourier criterion); $\mu_n = (2n-1)\frac{\pi}{2}$ are the roots of the characteristically equation.

It is convenient to express the amount of the substance ΔM diffusing into the body by the mean concentration of the medium that is easily determined by means of weightening $\Delta M = (\tilde{C} - C_0)M_0$, where, $\tilde{C}$ is the mean concentration of the medium in the body equal to:

$$\tilde{C}(t) = \frac{1}{2\pi}\int_{-R}^{R} C(x,t)dx \quad (1.28)$$

As a result of substitution of the value of C from equation (1.27) into subintegral expression (1.28), we get a law of change of mean concentration in time:

$$\frac{C_\infty - \tilde{C}}{C_\infty - C_0} = \sum_{n=1}^{\infty} B_n \exp(-\mu_n^2 F_0) \quad (1.29)$$

where, $B_n = 8/[(2n-1)^2\pi^2]$

The obtained result may be simplified if we will consider a very small ($F_0 < 0.2$) and a large ($F_0 < 0.2$) times of influence on the medium. So, for $F_0 < 0.2$ the following equation coincides well with the exact solution:

$$\tilde{\theta} = \frac{C_\infty - \tilde{C}}{C_\infty - C_0} = 1 - 2\sqrt{\frac{F_0}{\pi}} \quad (1.30)$$

or, the same

$$\frac{\tilde{C} - C_0}{C_\infty - C_0} = 2\sqrt{\frac{F_0}{\pi}} \quad (1.31)$$

On the other hand, for large values of Fourier criterion, practically, beginning with $F_0 < 0.2$, in calculations in expansion (1.31), we can assume that all the members of the series are small compared with the first one; then the calculation formula is simplified:

$$\tilde{\theta} = \frac{C_\infty - \tilde{C}}{C_\infty - C_0} = \frac{8}{\pi^2} \exp(-\frac{\pi^2}{4} F_0) \quad (1.32)$$

Under unbounded increase of time $C_\infty = C_n$, therefore in the state of complete saturation of the body that is attained already for $X_T = -\frac{1}{T}\nabla T$, the concentration of the diffusate is:

$$\tilde{C} = \frac{\Delta M_\infty}{M_0} = C_n = C_\infty \quad (1.33)$$

Thus, for determining the quantity $C_n = C_\infty$ of the material in the given medium, it is necessary to measure increase of the mass $\Delta M = M - M_0$ to the state of complete saturation of the sample; by attaining the final value M_∞ the quantity C_∞ is $\Delta M / M_0$.

Unbounded cylinder. The initial concentration of the medium $C_0 = 0$, concentration on the surface at time $t = 0$ becomes equal to C_∞. The concentration distribution is subjected to the law:

$$\theta = \frac{C_\infty - C(x,t)}{C_\infty - C_0} = \sum_{n=1}^{\infty} A_n J_0(\mu_n \frac{r}{R}) \exp(-\mu_n^2 \frac{Dt}{R^2}) \quad (1.34)$$

Where, $A_n = \frac{2}{\mu_n J_1(\mu_n)}$, μ_n are the roots of the characteristic equation and equal: $\mu_1 = 2.4048$; $\mu_2 = 5.5201$; $\mu_3 = 8.6537$; r,R is the current and the greatest value of the radius; J_0 is first kind Bessel function of zero order.

The solution is tabulated in (Stepanov and Shlenscy, 1981), the monograms for calculations may also be found in the solution. The mass absorption (the amount of diffused substance) equals:

$$\tilde{C}(t) = \frac{r}{R^2}\int_0^R rC(r,t)dr \tag{1.35}$$

After integration we get:

$$\tilde{\theta} = \frac{C_\infty - \tilde{C}(t)}{C_\infty - C_0} = \sum_{n=1}^{\infty} B_n \exp(-\mu_n^2 F_0) \tag{1.36}$$

It should be noted that the solution (1.36) is valid a plate, ball, and cylinder. For a ball $\mu_n = n\pi$, $B_n = \frac{6}{\mu_n^2}$.

Parallelepiped. In the general case, the three-dimensional diffusion process is realized on all three coordinate axis x, y, z when the medium contacts with all lateral bounds of a parallelepiped. Such a process is described by differential diffusion equation:

$$\frac{\partial C}{\partial t} = divgradC \tag{1.37}$$

If the initial concentration of the medium equals C_0 and the sizes of the parallelepiped are $2R_1 \times 2R_2 \times 2R_3$, then the solution of the problem on concentration distribution is in the form of a product of three solutions for unbounded plates that represent the borders of the parallelepiped:

$$\theta(x,y,z,t) = \theta_x(x,t)\cdot\theta_y(y,t)\cdot\theta_z(z,t) \tag{1.38}$$

We can write the similar product of three solutions for the total mass content as well:

$$\tilde{\theta} = \theta_x\theta_y\theta_z = \sum_{n=1}^{\infty}\sum_{m=1}^{\infty}\sum_{k=1}^{\infty} B_n B_m B_k \exp[-(\mu_n k_1^2 + \mu_m k_2^2 + \\ +\mu_k k_3^2)F_0] \tag{1.39}$$

where, the coefficients B and μ are calculated similar to B_n and μ_n in equation (1.29); $k_i = R_i/R$ $(i = 1,2,3)$; R is a generalized parameter defined by the formula:

$$\frac{1}{R^2} = \frac{1}{R_1^2} + \frac{1}{R_2^2} + \frac{1}{R_3^2} \tag{1.40}$$

A cylinder of finite length. The initial concentration C_0, medium's concentration on surface C_∞, cylinder's length 2ℓ, diameter $2R$.

We calculate the medium concentration distribution in the volume by finding the solutions of the problem for an unbounded cylinder and an unbounded plate:

$$\theta(r,z,t) = \theta_{ö}(r,t)\cdot\theta_{ïë}(z,t) \tag{1.41}$$

Accordingly, we calculate the mass content as well:

$$\tilde{\theta} = \tilde{\theta}_{ö}\tilde{\theta}_{ïë} = \sum_{n=1}^{\infty}\sum_{m=1}^{\infty} B_n B_m \exp[-(\mu_n^2 + \mu_m^2 k_\ell)F_0] \tag{1.42}$$

where, $B_n = 4/\mu_n^2$, $B_m = 4/\mu_m^2$, $k_\ell = R/\ell$, $F_0 = Dt/R^2$

Cite the examples of using the indicated relations.

Example 1 It is known (Stepanov and Shlenscy, 1981) that at normal pressure, the coefficient of water diffusion into polypropylene

$D = 4 \cdot 10^{-6}$ ñm^2/hr, $C_\infty = 0.4\%$, and under pressure (hydrostatic) 60 MPa $D_{60} = 2 \cdot 10^{-6}$ ñm^2/hr, $C_\infty = 0.3\%$.

Calculate how many times will be the process of diffusion of water into polypropylene moderated at pressure 60 MPa compared with the action of normal pressure if $C_\infty = 0$.

On the base of equation (1.30) for normal pressure $\tilde{C}/C_0 = 2\sqrt{F_0/\pi}$. For high pressure $\tilde{C}_{60}/C_{\infty 60} = 2\sqrt{F_{060}/\pi}$, where $F_{060} = D_{60}\, t_{60}/R^2$. Having divided one equality by the another one, we get:

$$\frac{\tilde{C}C_{\infty 60}}{C_\infty \tilde{C}_{60}} = \sqrt{\frac{F_0}{F_{060}}} = \sqrt{\frac{Dt}{D_{60} t_{60}}} \qquad (1.43)$$

Hence we define the time for attaining the equal value of the mean concentration $\tilde{C} = \tilde{C}_{60}$:

$$\frac{t_{60}}{t} = \frac{D}{D_{60}} \frac{C_\infty^2}{D_{\infty 60}^2} = \frac{4 \cdot 10^{-6} \cdot 0.4^2}{2 \cdot 10^{-6} \cdot 0.3^2} = 3.55 \qquad (1.44)$$

Thus, the pressure increase reduces to moderation of the diffusion process by increasing the water mass 3.55 times.

Example 2 Define the increment of the mass of a sample made of polypropylene in the form of a plate of dimensions $0.1 \times 10 \times 10$ cm^3 by absorbing water under normal pressure if $C_0 = 0$ and $t = 1$ days.

For a sample of dimensions $1/R_1^2 = \frac{1}{10^2} << \frac{1}{R_2^2} = 1/0.1^2$ we can use the formulae obtained for an unbounded plate.

We calculate beforehand the Fourier criteria using the data of the previous problem:

$$F_0 = \frac{Dt}{R^2} = \frac{4 \cdot 10^{-6} \cdot 24}{0.1^2} = 9.6 \cdot 10^{-3} \tag{1.45}$$

Since, $F_0 < 0.2$, formula (1.31) will be valid. As a result of substitution we get:

$$C = C_\infty \cdot 2\sqrt{F_0/\pi} = 0.4 \cdot 2\sqrt{\frac{9.6 \cdot 10^{-4}}{\pi}} = 4.4 \cdot 10^{-2}\% \tag{1.46}$$

The obtained result certifies on very negligible water absorption of a polypropylene and its well water-resistance.

Bounds of applicability of exact analytic solutions. The characteristics of water absorption and chemical resistance are defined on samples whose sizes are regulated by appropriate standards that define the form of the samples, test conditions, their duration. In some cases, while conducting tests, for providing one dimensionality of diffusion flux, the end face or lateral surfaces of samples are hydro-isolated. For processing the test results, these or other analytic solutions of mass-transfer problems are used. For defining diffusion characteristics of materials with the least error, on the samples whose form is often determined by the state of delivery of the material (pipe, sheets, beams, rods, etc.) one has to know which of the formulae will be used. For example, if the tested sample has the form of a beam of a rectangular section, it is necessary to decide for which sizes the simpler formulae are applicable for an unbounded domain, or for what sizes the more complicate formula will be used for a parallelepiped. Cite an example.

For water absorption testing of materials made of laminated plastics, the samples are cut in the form of a square with the side 50 ± 1mm and of thickness equal to the thickness of the sheet of the tested material. Obviously, for small thicknesses of isotropic samples, we may neglect the amount of liquid penetrating through the end faces compared with the amount of liquid penetrating through the large lateral surfaces of samples. In other words, the liquid flux may be considered as one-dimensional.

Using the above cited formulae, we can define the admissible thickness of the sample $2R_3$, under which with the given accuracy of calculations, the diffusion process may be considered one-dimensional, and the simpler formulae obtained for an unbounded plate may be used. For this use relation (1.40). For a square sample with the sides 50 mm $R_1 = R_2 = 2.5\text{cm}$. Then

$$1/R_1^2 + 1/R_2^2 = 1/2.5^2 + 1/2.5^2 = 0.32\text{cm}^2$$

The diffusion process will be considered one-dimensional if the following condition is satisfied:

$$1/R_1^2 + 1/R_2^2 + 1/R_3^2 = 1/R^2 \approx 4R_3^2$$

that is the addends $1/R_1^2 + 1/R_2^2$ are slightly small compared with the addend $1/R_3^2$ constitutes only several percentage of $1/R_3^2$. Prescribe such a share-accuracy of calculation is 2%. Then in the example under consideration:

$$1/R_1^2 + 1/R_2^2 = 0.02 \cdot 1/R_3^2 = 0.32\text{cm} = 3.2\text{mm}$$

Hence, $2R_3 = 5\text{mm}$, $R_3 = \sqrt{0.02/0.32} = 0.25\text{cm} = 2.5\text{mm}$. If the sample's thickness exceeds the found quantity, then for providing one-dimensional diffusion process, either the end faces of surface should be hydro-isolated or the sample's area should be extended.

Speaking on bounds of applicability of exact analytic solutions, it should be taken into account that all the above reduced solutions of mass-transfer problems are valid within the accepted assumptions, in particular, by fulfilling the Fick's law. In this connection, such an assumption must be verified for each specific pair, polymer-liquid. Naturally, it is necessary to be convinced that only one diffusion process but not several diffusion and chemical interactions processes proceed, and also that the quantity of diffusion coefficient is constant. The simplest method of the mentioned checking is to measure the mass content of the samples at various times and to construct the dependence $\Delta M/M_0 = f(t)$. If the Fick's law is fulfilled and the diffusion coefficient is a constant, then according to equation

(1.31), the curves of increase of the mass of samples in the coordinates $\Delta M/M_0 = C - \sqrt{t}$ at the initial area should be straight lines.

Another, more reliable checking is the investigation of medium concentration distribution on the cross section of samples. This may be done, for example, by an optic method, that is to observe the motion of front of medium of the given concentration or by separating the sample into separate layers by means of micro atom with the subsequent change of concentration of medium in these layers. The last operation may be executed by using radioactive isotopes (Sachs and Lubahn 1946, Siebel 1944, Stepanov and Shlenscy 1981).

Application of indicators changing the cooler in moving diffusing liquid gives fine results. As it follows from relation (1.24), the law of motion of the front of constant concentration medium is connected by a simple dependence with diffusion coefficient. It we want to get the diffusing medium concentration θ* that changes the cooler of the indicator, on (1.24) we get:

$$\theta^* = \frac{C_\infty - C}{C_\infty - C_0} = erf\left(\frac{x*}{2\sqrt{Dt}}\right) = const \qquad (1.47)$$

Hence, it follows that $\frac{x*}{2\sqrt{Dt}} = const$. Then the motion of the front of diffusing liquid is subjected to the relation:

$x^* = B2\sqrt{Dt}$ (1.48)that is, the depth of the front of liquid penetration x^* is proportional to the squared root from the medium influence time. If such a law is observed in the test, this means that the Ficks's law is fulfilled.

1.2 EXPERIMENTAL–THEORETICAL METHOD FOR DEFINING PHYSICO–MECHANICAL PROPERTIES OF POLYMER MATERIALS WITH REGARD TO INFLUENCE OF CORROSIVE LIQUID MEDIUM

In constructions made of polymer materials, and working in contact with various corrosive liquid and gassy media, their operational characteristics sharply deteriorate. This reduces to premature failure of constructions, meanwhile the serviceability of constructions without the medium's influence remains. However, account of influence of corrosive liquid media introduces some difficulties to calculation of structural elements made of polymer, especially of composite materials, since the influence of corrosive liquid and gassy media is not restricted by the surface action but it is of volume character. The gassy and liquid media intensively diffuse into the internal layers of the material change their chemical composition causing swelling and change of physico–chemical properties of polymer and composite materials. Effect of influence of corrosive liquid media happens in the form of physical and chemical action (Eroshenko and Zaychik, 1984; Eroshenko et. al., 1980; Ferry, 1963; Kargin and Slonimsky, 1960; Shen et. al., 1976; Siebel, 1944; Tikhomirov, 1970; Urzhumtseva, 1982; Zuev, 1972).

The first one reduces to reversible changes in the materials structure (swelling, dissolving) that vanish after removing the medium. The second one reduces to irreversible changes in the structure of polymers (Aliyev, 1995, 1998; Aliyev and Gabibov, 1994; Sachs and Lubahn, 1946; Stepanov and Shlenscy, 1981).

It should be also noted that because of diffusion, gas formations, chemically active ions and free electrons are washed away in certain degree from the micropores of polymers.

In connection with the above-stated ones there remains open a question: how does the physico–chemical properties of polymer materials and also a problem of mechanics of deformable solids (strength, stability, vibration, failure) depend on their physico–chemical changes. In particular, what is the quality and quantity character of dependence of serviceability of different constructions made of polymer and composite materials and also of their mechanical properties on their contact with corrosive liquid and gassy media. The exact solution of such a problem that is, the joint solution of diffusion and continuum mechanics is a very difficult mathematical problem of connected type. At the same time, such an approach requires to know exact values of initial properties of the material, of diffusion coefficient and boundary conditions of the problem. Otherwise, the mathematical problem may be ill-posed. Hence it is seen that the problem under consideration may get its fundamental solution by investigating it on the joint of two sciences: physico–mathematical and physico–chemical theories.

Below we suggest an experimental-theoretical method on definition of physico–mechanical characteristics of polymer and composite materials with regard to change of their physico–chemical properties, and also suggest concrete models of mechanical deformation (Aliyev, 1984, 1995, 1998, 2012, 2012, Aliyev and Gabibov, 1994).

Let the given samples made of polymer materials and having initial sizes a_0, b_0, ℓ_0 stay in the corrosive liquid medium $t_1, t_2, t_3, ..., t_n$ time. Under the action of the corrosive medium they will change their geometrical sizes and physico–chemical properties whose account will be accepted in the form of a swelling function in time, in the form (Aliyev, 2012, 2012):

$$\lambda(t) = \frac{Q(t) - Q_0}{Q_0} \tag{1.49}$$

where: $Q(t) = \gamma(t) \cdot V(t) = \gamma(t) \cdot a(t) \cdot b(t) \cdot \ell(t)$ and $Q_0(t) = \gamma_0 \cdot V_0 = \gamma_0 \cdot a_0 \cdot b_0 \cdot \ell_0$ are current and initial weights of the samples at moments $t > 0$ and $t = 0$. We construct the diagrams:

$$\lambda(t) = \frac{Q(t) - Q_0}{Q_0} = f(t) \quad \varepsilon_V^0 = \frac{V(t) - V_0}{V_0} = f_V(\lambda),$$

$$\varepsilon_\ell^0(\lambda) = \frac{\Delta\ell(\lambda)}{\ell_0} = \frac{\ell(t) - \ell_0}{\ell_0} = f_l(\lambda),$$

$$\varepsilon_F^0(\lambda) = \frac{\Delta F^0}{F_0} = \frac{a(\lambda)b(\lambda) - a_0 b_0}{a_0 b_0} = f_F(\lambda) \tag{1.50}$$

where, $\varepsilon_V^0(\lambda)$, $\varepsilon_\ell^0(\lambda)$, $\varepsilon_F^0(\lambda)$ are cubic, longitudinal and lateral deformations of the sample at time $t > 0$.

A group of thermoplastic materials widely applied oilfield practice: polyethylene of lower density (PELD), polyethylene of high density (PEHD), polypropylene (PP), shock resistance polystyrene

(PS), polyamid-610 (PA), and also polyamide (Figure 1.1, Table 1.1) were used in place of experimental investigation.

Oil with dynamic viscosity $\mu = 32.41 \cdot 10^{-3}$ Pa sec and with 0.277% of asphalthene acids was taken in place of corrosive liquid medium in which the samples were maintained. Experimental investigations were conducted by us at the Institute of Mathematics and Mechanics of the National Academy of Sciences of Azerbaijan and Oil Academy of the Republic of Azerbaijan.

Essential influence of oil corrosiveness on change of mechanical characteristics and physico–chemical properties of polymers was experimentally established. The samples made of polymer materials maintained in oily medium five weeks have swelling from 2 to 10%. Therewith, the cubic deformation constitutes from 2% to 10%, longitudinal deformation from 0.6 to 3%, change of the cross section area from 1% to 7%. Linear dependence of coordinate deformations on the swelling parameter λ for small deformations, that is, for $\varepsilon_x^2 << \varepsilon_x$ is typical for the tested class of polymer material

$$\varepsilon = \alpha\lambda, \; \varepsilon_V = 3\alpha\lambda \tag{1.51}$$

where, α is a linear coefficient of longitudinal swelling introduced by us.

For the considered class of polymer material whose samples were maintained in the oily medium, we first found experimentally the linear coefficients of longitudinal swelling α (Table 1.2). Knowing the experimental dependences:

$$\lambda(t) = \frac{Q(t) - Q_0}{Q_0} = \frac{\gamma(t)V(t) - \gamma_0 V_0}{\gamma_0 V_0}, \; \varepsilon_\ell(\lambda) = \alpha\lambda,$$

$$\varepsilon_V(\lambda) = 3\alpha\lambda \tag{1.52}$$

Changeability of the density $\gamma(\lambda)$ of polymer materials, depending on swelling parameter λ for small deformations will have the following form:

$$\frac{\gamma(t)}{\gamma_0} = \frac{1+\lambda}{1+3\alpha\lambda} \tag{1.53}$$

For $\alpha = 0$, the density changes at the expense of chemical transformation of polymer material when it contacts with corrosive medium without changing the volume. The case $\alpha = 0$ and $\gamma(t) = \gamma_0$ correspond to the case when there is no influence of corrosive medium. It is shown in Figure 1.2 that the density of the polymer material $\gamma(t)$ in the corrosive oily medium also changes. Moreover, for one class of polymers the density falls by the quantity from 0.1 to 0.5%, but for another class increases to 2%.

According to the results of experiments and formula (1.53), it is seen that with increasing maintenance of polymer material in corrosive medium, on one hand volume extension of the material and increase of the weight of the body occurs, but on the other hand, the density for one materials decreases, for another one increases its value. On this basis, we can deduce that depending on the kind of polymer material and type of corrosive liquid medium, the density change function may increase or decrease depending on degree of swelling parameter λ

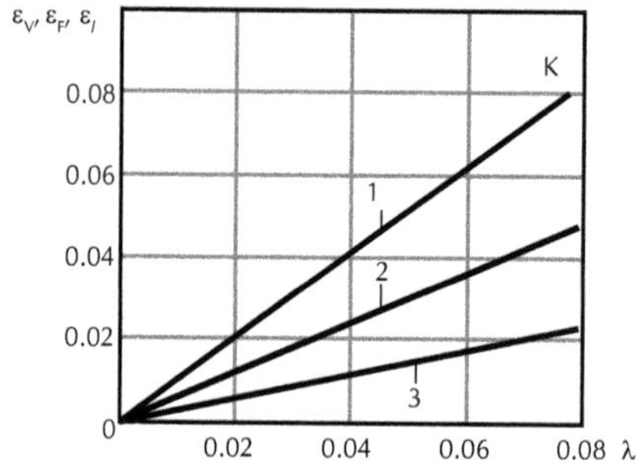

a

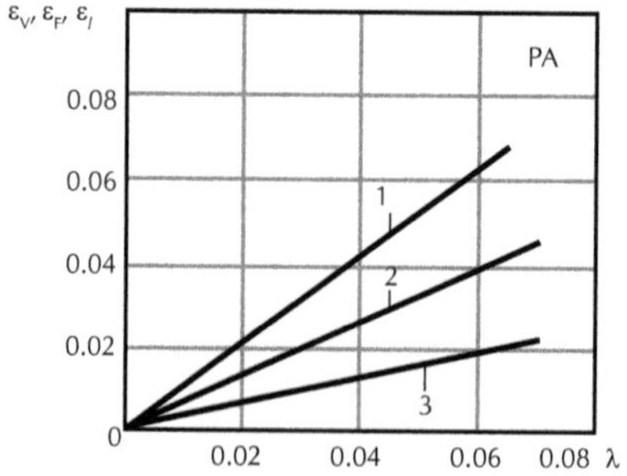

b

FIGURE 1.1 *(Continued)*

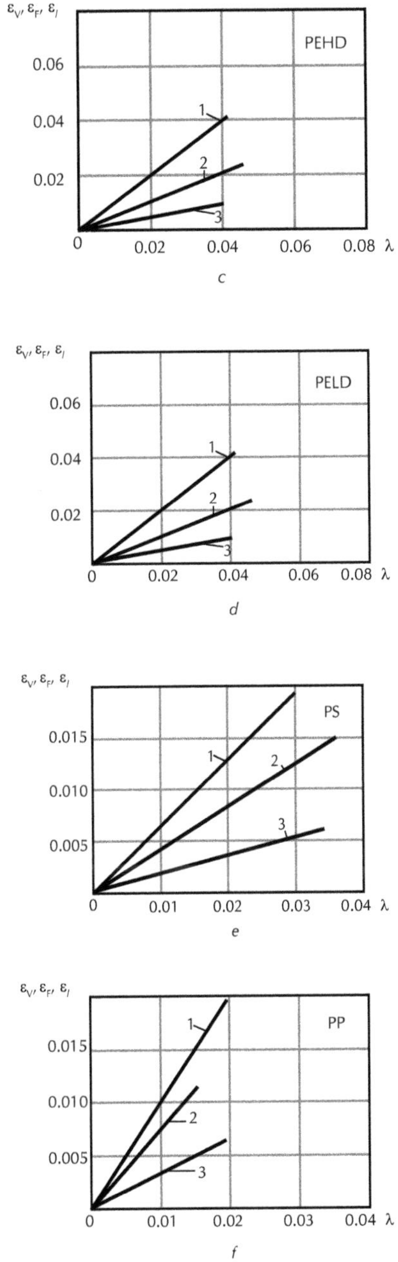

FIGURE 1.1 Dependence of longitudinal $\varepsilon_\ell(\lambda)$, lateral $\varepsilon_F(\lambda)$ and volume $\varepsilon_V(\lambda)$ deformations on swelling parameter λ for polymer materials: a). K; b). PA; v). PEHD; q). PELD; d). PS; e). PP

TABLE 1.1 Dependence of ε_ℓ^0, $\varepsilon_\perp$, ε_V^0 on swelling parameter λ

Material	Day t	$Q(t)$ (gf)	$\lambda(t)$	$l_0(t)$ (mm)	$a_0(t)$ (mm)	$b_0(t)$ (mm)	F (t) (mm²)	ε_ℓ^0	$\varepsilon_\perp$	ε_V^0
1	2	3	4	5	6	7	8	9	10	11
K	0	0.3264	0	25.00	1.95	5.94	11.5830	0	0	0
	1	0.3219	0.0168	25.15	1.96	5.96	11.6816	0.0060	0.0090	0.0145
	2	0.3371	0.0328	25.36	1.97	5.97	11.7609	0.0144	0.0116	0.0299
	3	0.3409	0.0444	25.48	1.98	5.98	11.8404	0.0152	0.0223	0.0418
	4	0.3471	0.0634	25.64	2.00	5.99	11.9800	0.0256	0.0310	0.0607
	5	0.3535	0.0830	25.75	2.04	6.05	11.3420	0.0300	0.0690	0.0978
PA	0	0.3191	0	25.00	1.95	5.95	11.6025	0	0	0
	1	0.3228	0.0116	25.12	1.96	5.96	11.6816	0.0048	0.0068	0.0116
	2	0.3270	0.0248	25.18	1.97	5.97	11.7609	0.0112	0.0136	0.0181
	3	0.3309	0.0370	25.38	1.98	5.98	11.8464	0.0152	0.0205	0.0369
	4	0.3349	0.0495	25.54	1.99	5.99	11.9201	0.0216	0.0274	0.0459
	5	0.3397	0.0646	25.68	2.00	6.06	11.1200	0.0252	0.0446	0.0730
PEHD	0	0.2774	0	25.0	1.96	5.96	11.6816	0	0	0
	1	0.2797	0.0083	25.08	1.97	5.96	11.7412	0.0032	0.0051	0.0083
	2	0.2808	0.0123	25.14	1.97	5.98	11.7609	0.0056	0.0067	0.0123
	3	0.2851	0.0279	25.12	1.98	5.98	11.8404	0.0088	0.0136	0.0122
	4	0.2877	0.0371	25.28	1.99	6.00	11.9400	0.0112	0.0221	0.0335
	5	0.2902	0.0462	25.39	1.99	6.05	11.0395	0.0156	0.0306	0.0470

TABLE 1.1 *(Continued)*

PELD	0	0.2664	0	25.0	1.95	5.94	11.5830	0	0	0
	1	0.2695	0.116	25.12	1.96	5.95	11.6620	0.0048	0.0068	0.0116
	2	0.2725	0.0229	25.26	1.97	5.96	11.7412	0.0104	0.0136	0.0242
	3	0.2741	0.0291	25.33	1.97	5.98	11.7609	0.0132	0.0154	0.0288
	4	0.2767	0.0387	25.40	1.98	5.98	11.8404	0.0160	0.0222	0.0386
	5	0.2808	0.0520	25.52	2.00	6.02	11.0400	0.0208	0.0395	0.0611
PS	0	0.3082	0	25.0	1.97	5.96	11.7412	0	0	0
	1	0.3105	0.0074	25.5	1.97	5.96	11.7412	0.0020	0.0001	0.0020
	2	0.3129	0.0154	25.10	1.98	5.97	11.8206	0.0040	0.0067	0.0107
	3	0.3147	0.0210	25.12	1.98	5.97	11.8206	0.0048	0.0067	0.0116
	4	0.3170	0.0285	25.16	1.99	5.98	11.9002	0.0064	0.0135	0.0200
	5	0.3193	0.0316	25.20	1.99	5.99	11.9201	0.0080	0.152	0.0234
PP	0	0.2628	0	25.0	1.96	5.96	11.6816	0	0	0
	1	0.2637	0.0035	25.03	1.96	5.96	11.6816	0.0010	0.0001	0.0012
	2	0.2655	0.0103	25.08	1.97	5.97	11.7609	0.0032	0.0068	0.0100
	3	0.2667	0.0148	25.13	1.97	5.97	11.7609	0.0052	0.0068	0.0180
	4	0.2681	0.0202	25.16	1.98	5.98	11.8404	0.0060	0.0136	0.0201
	5	0.2697	0.262	25.18	1.98	5.99	11.7602	0.0061	0.0153	0.0226

TABLE 1.2 Coefficients of linear α and volume α_v swelling of polymer material under the action of corrosive oily medium

The polymer's brand	α	$J_i = \sum^{n} L_{ik} X_k$	a_F
PA	0.3333	1.0000	0.6667
PEHD	0.3618	1.0854	0.7236
PELD	0.3452	1.0357	0.6905
PS	0.2333	0.7000	0.4667
PP	0.3000	0.9000	0.6000
K	0.3368	1.0103	0.6736

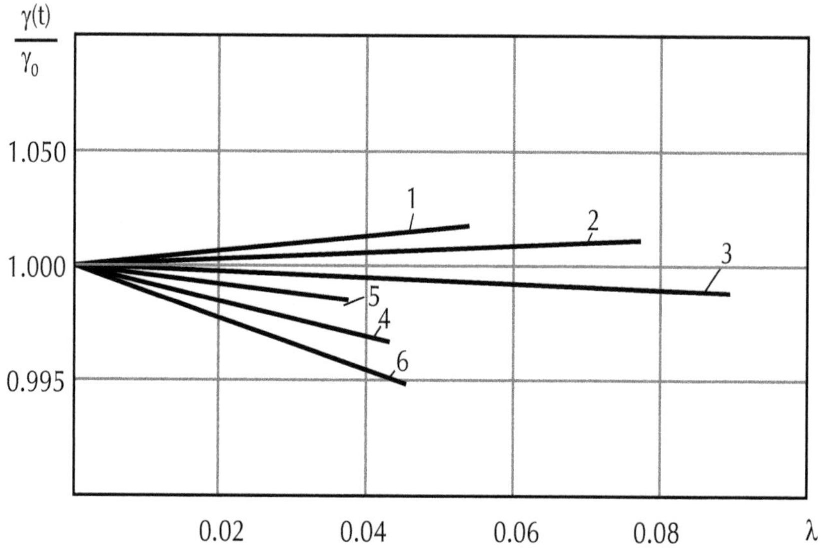

FIGURE 1.2 Dependence of change of density of polymer material $\frac{\gamma(t)}{\gamma_0}$ on swelling parameter λ for: 1—PS; 2—PA; 3—K; 4—PP; 5—PEHD; 6—PELD

Along the above stated we note the following. According to Table 1.1 in Figure 1.3 for a class polymer material the dependence of the swelling parameter $\lambda(t)=\frac{Q(t)-Q_0}{Q_0}$ on time t is represented in the form:

$$\lambda(t)=qt \quad (1.54)$$

Here the coefficient q of dimension $\left(\frac{1}{\text{day}}\right)$ equals:

$$q=\frac{\lambda(t_n)}{t_n}=const \quad (1.55)$$

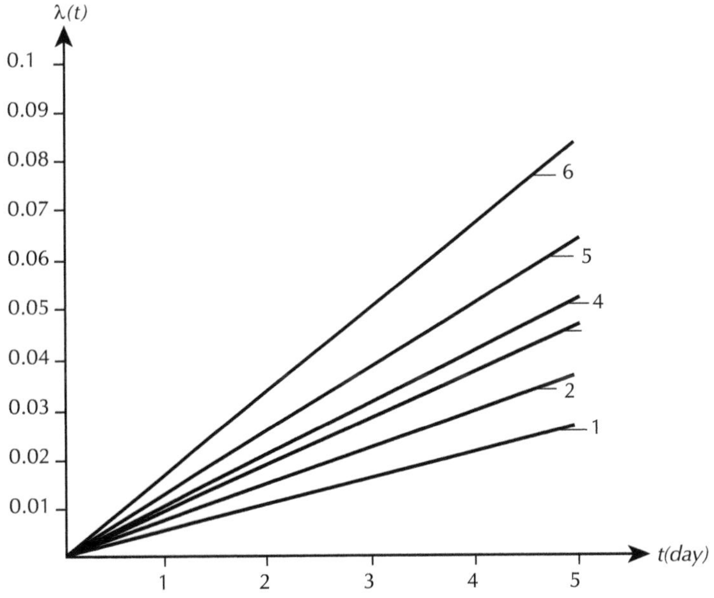

FIGURE 1.3 Dependence of the swelling parameter $\lambda(t)=\frac{Q(t)-Q_0}{Q_0}$ on time t for the following polymer materials: 1—PP; 2—PS; 3—PEHD; 4—PELD; 5—PA; 6—K

Numerical values of the dependence λ(t) on time t for a series of polymer materials is represented below (see Table 1.3):

for PP: $\lambda(t)=0.00524t$; for PS: $\lambda(t)=0.00722t$; for PEHD: $\lambda(t)=0.00924t$; for PELD: $\lambda(t)=0.0104t$; for PA: $\lambda(t)=0.01292t$.

While changing the dimensionless parameter $\tilde{\lambda}=\frac{\lambda}{\lambda_{max}}$ in the interval $0\le\tilde{\lambda}\le1$ its appropriate value of time t will change in the following interval:

$$0\le t\le\frac{\lambda_{n\,max}}{q} \tag{1.56}$$

TABLE 1.3 Numerical values of the dependence $\lambda(t)$ on time t for a series of polymer materials

Materials brand	$\lambda(t)=tq$	λ_{max}	$0\le\tilde{\lambda}\le1$	$0\le t\le\frac{\lambda_{max}}{q}$ (*day*)
PP	$\lambda(t)=0.00524t$	0.02	$0\le\tilde{\lambda}\le1$	$0\le t\le3.8168$
PS	$\lambda(t)=0.00722t$	0.02	$0\le\tilde{\lambda}\le1$	$0\le t\le2.7701$
PEHD	$\lambda(t)=0.00924t$	0.0275	$0\le\tilde{\lambda}\le1$	$0\le t\le2.9762$
PELD	$\lambda(t)=0.0104t$	0.035	$0\le\tilde{\lambda}\le1$	$0\le t\le3.3654$
PA	$\lambda(t)=0.01292t$	0.028	$0\le\tilde{\lambda}\le1$	$0\le t\le2.1672$

In other words, physico–mechanical properties of the polymer material $E(\lambda)$, $G(\lambda)$, $\nu(\lambda)$ under the action of corrosive liquid medium will change in the course of time $0\le t\le\frac{\lambda_{max}}{q}$. For the times $t\ge\frac{\lambda_{max}}{q}$ mechanical properties of polymer material stop to change and accept constant values, that is, the corrosive medium doesn't influence on changeability of physico–mechanical properties of polymer material. For example, the material PP changes its features in the course of 3.8168 days; PS in the course of

2.7701 days; PEHD in the course of 2.9762 day; PELD in the course of 3.3654 days; PA in the course of 2.1672 days (see Table 1.3).

Note the following peculiarity. Let the samples made of two different polymer materials stay equal time in aggressive medium. During this time each of the polymer materials change their physico–mechanical properties because of influence of corrosive liquid and their swelling parameters will be different. Show the form of mathematical relation between the swelling parameters λ_1 and λ_2. According to formula (1.54), this relation will be of the form [9,12]:

$$\lambda_2 = \frac{q_2}{q_1}\lambda_1 \tag{1.57}$$

Hence it follows that if the swelling parameter λ_1 for the first polymer material changes in the interval $0 \le \lambda_1 \le \lambda_{1\max}$, that is, for $0 \le \tilde{\lambda}_1 = \frac{\lambda_1}{\lambda_{1\max}} \le 1$, then the swelling parameter λ_2 of the second polymer material will change in the interval $0 \le \lambda_2 = \frac{q_2}{q_1}\lambda_1 \le \lambda_2^* \le \lambda_{2\max}$, that is, for $0 \le \tilde{\lambda}_2 \le \frac{\lambda_2^*}{\lambda_{2\max}} \le 1$. Account of such an effect is very necessary by solving the problems of stability and vibration of laminated constructions made of polymer materials whose layers possess different physico–chemical properties. In other words, it is necessary to use relation (1.57) for finding a unique solution of such class problems of mechanics.

Example 3: Let polymer materials of brands PP and PS with the following swelling properties be given: $q_1 = 0.00524$ and $q_2 = 0.00722$. Then the relation between their swelling parameters will be in the form:

$$\lambda_2 = \frac{0.00722}{0.00524}\lambda_1 = 1.3779\lambda_1 \tag{1.58}$$

Hence, it follows that at one and the same staying time of polymer materials in corrosive medium, to the numerical value of the swelling parameter λ_1 of the material of the brand PS there will correspond the numerical value of the swelling parameter λ_2 of the brand PP equal $\lambda_2 = 1.3779\lambda_1$.

1.2.1 LINEAR ELASTIC DEFORMATION MODEL OF POLYMER MATERIALS WITH REGARD TO CHANGE OF PHYSICO–CHEMICAL PROPERTIES

Polymer samples maintained in the corrosive liquid, for each fixed value of the parameter λ_n are tested for one-dimensional extension, and dependences $\sigma = f(\varepsilon)$ and dependence of ratio of lateral deformation $\varepsilon_\perp$ to longitudinal deformation ε_ℓ are constructed in the form (Aliyev, 1995, 2011, 2012):

$$\sigma = \frac{P}{F(\lambda)} = f(\varepsilon),\ \nu(\lambda) = -\frac{\varepsilon_\perp}{\varepsilon_\ell},\ \varepsilon = \frac{\Delta\ell(\lambda)}{\ell_0} \tag{1.59}$$

Here $F(\lambda)$, $\Delta\ell(\lambda) = \ell(\lambda) - \ell_0$ is the area of cross-section of the sample and absolute lengthening taken for each value of the parameter λ_n (Figure 1.3); the Poisson ratio $\nu(\lambda)$ is the ratio of lateral deformation $\varepsilon_\perp$ to longitudinal ε_ℓ; ε_ℓ^0 is the change of the length of the sample because of only swelling effect. In this case, the general longitudinal deformation ε of the sample is composed of deformation of only swelling $\varepsilon_\ell^0 = \alpha\lambda$ and deformation arising from mechanically applied load $\varepsilon_\ell(P)$ in the form:

$$\varepsilon = \varepsilon_\ell^0(\lambda) + \varepsilon_1(P) \tag{1.60}$$

We will assume that the dependence of deformation $\varepsilon_\ell(P)$ on the stress σ is linear:

$$\varepsilon_1(P) = \frac{\sigma}{E(\lambda)} \tag{1.61}$$

Substituting (1.61) and $\varepsilon_\ell^0 = \alpha\lambda$ in (1.60), we get:

$$\varepsilon = \frac{\sigma}{E(\lambda)} + \alpha\lambda \tag{1.62}$$

or

$$\sigma = E(\lambda)(\varepsilon - \alpha\lambda) \tag{1.63}$$

For σ = 0 from formula (1.63) we will have: $\varepsilon = \varepsilon_{\ell}^{0}(\lambda) = \alpha\lambda$.

This will correspond to the case of free extension, that is, swelling of the sample only under the action of corrosive liquid medium (Figure 1.4). The case

$$\varepsilon = 0\,,\ \sigma = -E(\lambda)\alpha\lambda \tag{1.64}$$

will correspond to the case of fastened sample at fixed ends and boundedness of its longitudinal deformation.

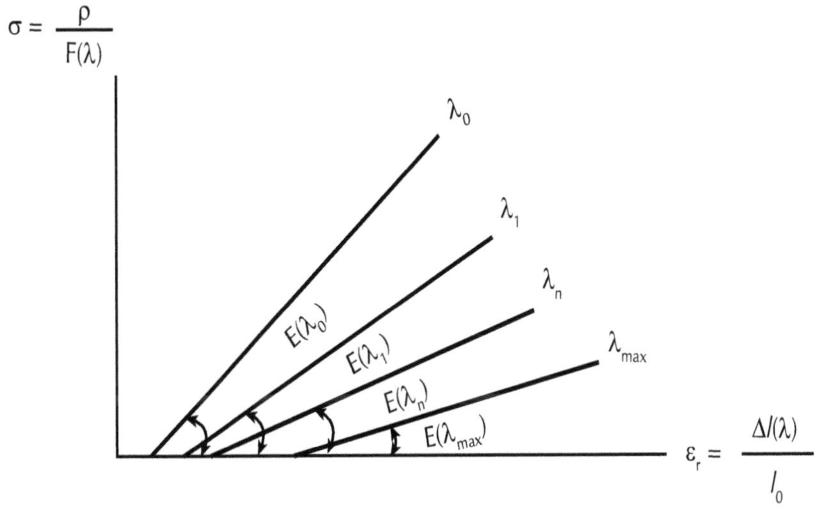

FIGURE 1.4 Diagram $\sigma = f(\varepsilon)$ for different values of swelling parameter λ_n of polymer material

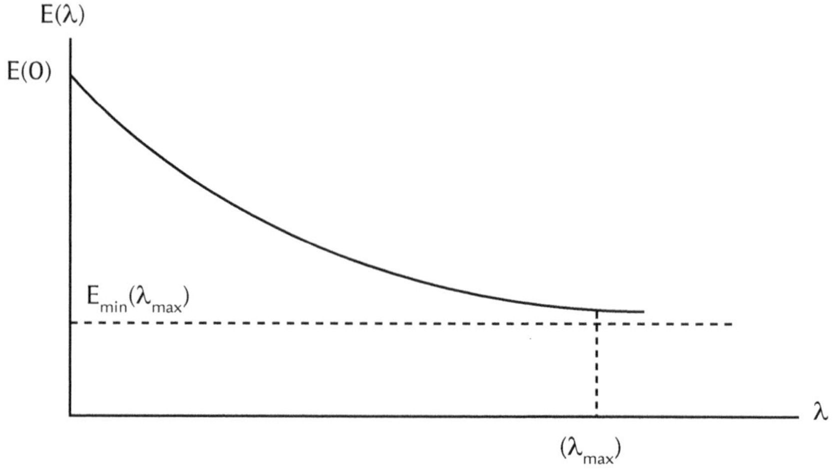

FIGURE 1.5 Dependence of elasticity modulus $E(\lambda)$ on the swelling parameter λ of polymer material.

Under the action of only corrosive medium, in the sample there will arise thrust stress of size $\sigma = -E(\lambda)\varepsilon_{\ell}^{0}(\lambda)$. Taking into account $\sigma = \frac{P}{F_0(1+\varepsilon_F)}$, the arising thrust compression force $P(\lambda)$ will equal:

$$P(\lambda) = \sigma F(\lambda) = -E\lambda F_0(1+\varepsilon_F)\alpha\lambda \tag{1.65}$$

Model (1.63) will be completely defined if the dependence of the elasticity modulus $E(\lambda)$ and the Poisson ratio $\nu(\lambda)$ on the swelling parameter λ (Figure 1.5) will be established. For that the following brands of polymer materials were tested for one-dimensional extension: PA, PELD, PEHD, PS, PP maintained in oily medium of dynamic viscosity $\mu = 32.41 \cdot 10^{-3}$ Pas and with 0.277% asphalt thene acids. The samples in the form of strips of width 6.0mm, of thickness 2.0mm, of length 25 mm (Figure 1.6, Table 1.4) were tested.

It is experimentally established that the dependence $\sigma \sim \varepsilon$ for polymer materials will be linear in $0 \le \lambda \le \lambda_{max}$ interval of change of swelling parameter λ. Secondly, it is seen from Figure 1.6 that the elasticity modulus $E(\lambda)$ falls by the exponential law $E(\lambda) = \alpha \exp(\rho\lambda)$ by 25% depending on degree of the swelling parameter λ, and the Poisson ratio $\nu(\lambda)$ increases by the weak linear law by 15% (Figure 1.7).

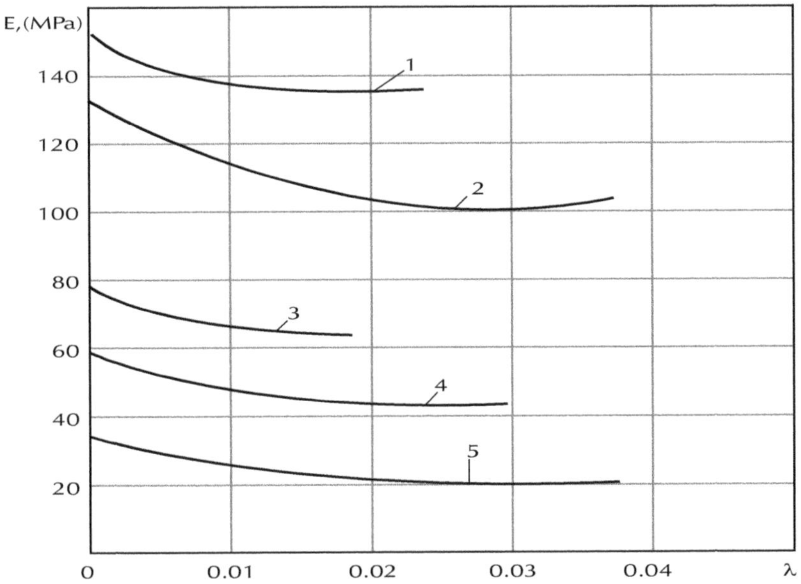

FIGURE 1.6 Dependence of elasticity modulus $E(\lambda)$ on swelling parameter λ for different plastics: 1—PS; 2—PA; 3—PP; 4—PEHD; 5—PELD

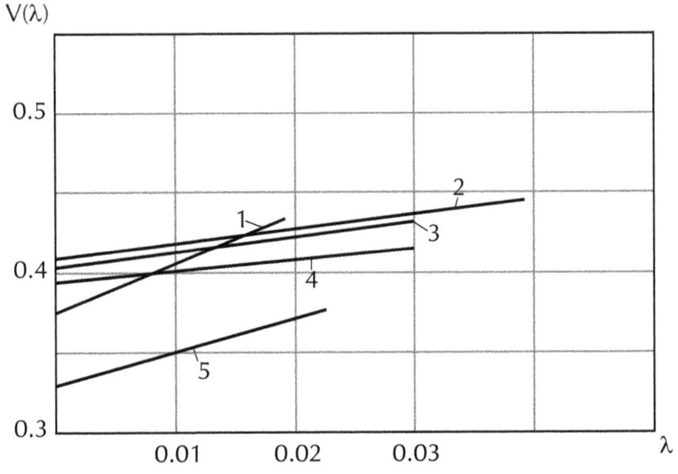

FIGURE 1.7 Dependence of Poisson ratio $\nu(\lambda)$ on swelling parameter λ for different plastics: 1—PP; 2—PA; 3—PELD; 4—PEHD; 5—PS

Knowing the experimental values $E(\lambda)|_{\lambda=0} = E_0$ and $E(\lambda)|_{\lambda=\lambda\max} = E_{\min}$, $\nu(\lambda)|_{\lambda=0} = \nu_0$, $\nu(\lambda)|_{\lambda=\lambda\max} = \nu_{\max}$ we approximate experimental curves $E(\lambda) = f_E(\lambda)$ and $\nu(\lambda) = f_\nu(\lambda)$, in the form:

$$E(\lambda) = E_0 (\frac{E_{\min}}{E_0})^{\lambda/\lambda_{\max}} = E_0 (\frac{E_{\min}}{E_0})^{\tilde{\lambda}} = E_0 b^{\tilde{\lambda}} \quad (1.66)$$

$$\nu(\lambda) = \nu_0 + (\nu_{\max} - \nu_0)\frac{\lambda}{\lambda_{\max}} = \nu_0 + k\tilde{\lambda} \quad (1.67)$$

Here, the dimensionless parameter $\tilde{\lambda} = \frac{\lambda}{\lambda_{\max}}$ changes in the interval $0 \le \tilde{\lambda} \le 1$. Substituting (1.66) in (1.63), an one-dimensional model of linear deformation of polymer material with regard to its physico–chemical property will take the form:

$$\sigma(\varepsilon, \lambda) = E_0 b^{\tilde{\lambda}} (\varepsilon - \alpha\lambda) \quad (1.68)$$

Here $b^{\tilde{\lambda}} = (\frac{E_{\min}}{E_0})^{\tilde{\lambda}}$ is a correction factor in the interval $0 \le b^{\tilde{\lambda}} \le 1$. The case $\tilde{\lambda} = 1$, $E_{\min} = 0$ corresponds to total decay of polymer material, that is to disappearance of molecular interaction forces. For $\tilde{\lambda} = 0$, $E_{\min} = E_0$ we get a model of deformation of polymer material under the action of only mechanical forces. In the case of a sample fastened by two ends and boundedness of its lateral deformation, that is, for $\varepsilon_\ell = 0$, the arising thrust stress and force depending on the swelling parameter λ, according to (1.68) will be equal to (1.68):

$$\sigma(\lambda) = -E_0 b^{\tilde{\lambda}} \cdot \alpha\lambda \quad (1.69)$$

$$P(\lambda) = \sigma(\lambda) \cdot F_0 (1 + \varepsilon_F) = E_0 F_0 b^{\tilde{\lambda}} (1 + \varepsilon_F(\lambda)) \cdot \alpha\lambda \quad (1.70)$$

We have experimentally established that the proportionality limit $\sigma_{pr.}(\lambda)$ for the studied class of polymer material falls by exponential law depending on the swelling parameter λ (Figure 1.8).

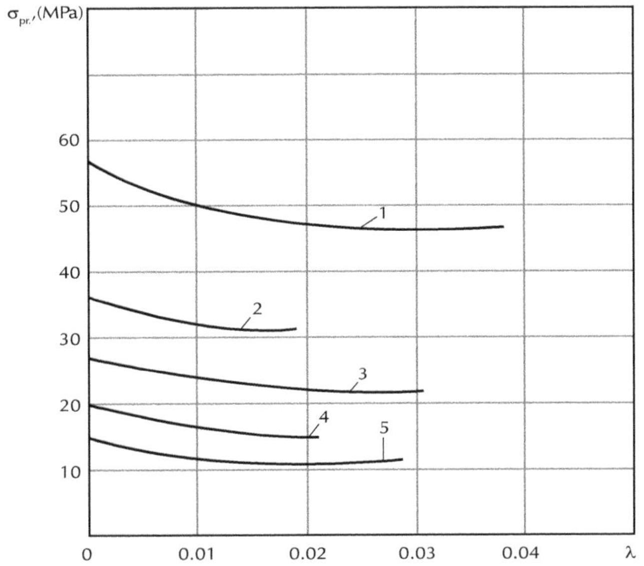

FIGURE 1.8 Dependence of proportionality limit $\sigma_{pr.}$ on swelling parameter λ for different plastics: 1—PA; 2—PP; 3—PEHD; 4—PS; 5—PELD

Summing up the above stated one, we can say that if it is required to calculate the quantities of stress and force in a one-dimensional polymer sample in the λ corrosive medium, then formulae (1.62)–(1.65) should be used. But if the asymptotic curve $E(\lambda) \sim \lambda$ is known, then formulae (1.68)–(1.70) may be used.

Example 4: Let a thin strip made of polymer material of geometrical sizes $a_0 = 6\text{mm}$, $b_0 = 2\text{mm}$, $\ell_0 = 25\text{mm}$ be in the corrosive oil medium with dynamic viscosity $\mu = 32.41 \cdot 10^{-3}$ Pa s and with 0.277% asphaltene acid. The samples were made of polymer materials of brands PA, PELD, PEHD, PS, PP.

Determine the quantities of thrust stress and forces in the strips depending only on the swelling parameter λ after their 5 day stay in corrosive medium. As the sample was fastened by two ends, its longitudinal deformation will equal to zero $\varepsilon_\ell = 0$. In this case, we use formulae (1.64) and (1.65), that is:

$$\sigma = -E(\lambda) \cdot \alpha\lambda \tag{1.71}$$

$$p(\lambda) = -E(\lambda) \cdot F_0(1+\varepsilon_F(\lambda)) \cdot \alpha\lambda \tag{1.72}$$

Using the experimental data of Tables 1.1 and 1.2, the appropriate calculated number values for $\frac{\sigma(\lambda)}{\sigma_{pr.}}$ and $P(\lambda)$ are given in Table 1.4.

TABLE 1.4 Calculated number values for $\frac{\sigma(\lambda)}{\sigma_{pr.}}$ and $P(\lambda)$

Material brand	$\varepsilon_\ell^0(\lambda)$	$F_0(1+\varepsilon_F(\lambda))$ (cm²)	$E(\lambda)$ (MPa)	$\sigma(\lambda)$ (MPa)	$P(\lambda)$ (kgf)	$\frac{\sigma(\lambda)}{\sigma_{pr.}} \cdot 100\%$
PA	0.0252	0.1212	101.5	-2.5578	-31	5.4
PELD	0.0208	0.1204	20.3	-0.4223	-5.0	3.7
PEHD	0.0156	0.1204	44.0	-0.6	-7.2	2.6
PS	0.0080	0.1192	125	-1	-11.9	5.9
PP	0.0061	0.1186	66.25	-0.4041	-4.8	1.2

It is seen from Table 1.4 that influence of corrosive liquid on rise of thrust stress composes 6% from the admissible value of material's stress. This value should be taken into account in the quantity of the admissible value of stress. It should be noted that values of thrust forces will be commeasurable with the critical values of stability of strips under compression. It follows from this example that in the considered case, that is, when a polymer strip is in corrosive medium, the strip may be strength but not stable. This says on necessity of investigation of stability problem of a strip situated under the action of corrosive liquid medium.

1.2.2 HOOKE'S GENERALIZED LAW FOR POLYMER MATERIAL WITH REGARD TO INFLUENCE OF CHANGE OF PHYSICO–CHEMICAL PROPERTIES

Below we suggest a three-dimensional linear-elastic model of deformation of polymer materials working in corrosive liquid media (Aliyev, 1995, 1998, 2012, 2012). Because of diffusion of corrosive liquid into polymer material and its physico–chemical changes, the total deformation of the body in three orthogonal directions will be composed of linear constituents of deformation of only swelling $\varepsilon_{\ell}^{0}(\lambda)$ and deformation of mechanical loadings ε(p) in the form:

$$\varepsilon_1 = \varepsilon_1(P) + \varepsilon_1^0(\lambda),\ \varepsilon_2 = \varepsilon_2(P) + \varepsilon_2^0(\lambda),$$

$$\varepsilon_3 = \varepsilon_3(P) + \varepsilon_3^0(\lambda) \tag{1.73}$$

where, $\varepsilon_1, \varepsilon_2, \varepsilon_3$ are total deformations. By linear dependence $\sigma \sim \varepsilon$ for an arbitrary value of the swelling parameter λ, the Hook's generalized law will be in the form:

$$\begin{cases} \varepsilon_1(P) = \dfrac{1}{E(\lambda)}[\sigma_1 - \nu(\lambda)(\sigma_2 + \sigma_3)] \\ \varepsilon_2(P) = \dfrac{1}{E(\lambda)}[\sigma_2 - \nu(\lambda)(\sigma_1 + \sigma_3)] \\ \varepsilon_3(P) = \dfrac{1}{E(\lambda)}[\sigma_3 - \nu(\lambda)(\sigma_1 + \sigma_2)] \end{cases} \tag{1.74}$$

Since for a homogeneous isotropic body the swelling effect is the same in all the directions, then there will appear only normal deformations $\varepsilon_1^0(\lambda), \varepsilon_2^0(\lambda), \varepsilon_3^0(\lambda)$, the tangential deformations will equal zero. Moreover:

$$\varepsilon_1^0(\lambda) = \varepsilon_2^0(\lambda) = \varepsilon_3^0(\lambda) = \frac{1}{3}\varepsilon_V^0 \tag{1.75}$$

Here $\varepsilon_V^0 = 3\alpha\lambda$ is the cubic deformation arising only because of swelling of the sample. Substituting the value of $\varepsilon_\ell^0 = \alpha\lambda$ and (1.75) into (1.73) and taking account that the sum of normal stresses equals $\sigma = \sigma_1 + \sigma_2 + \sigma_3$, we get:

$$\begin{cases} \varepsilon_1 = \dfrac{1}{E(\lambda)}[(1+\nu(\lambda))\sigma_1 - \nu(\lambda)\sigma] + \alpha\lambda \\ \varepsilon_2 = \dfrac{1}{E(\lambda)}[(1+\nu(\lambda))\sigma_2 - \nu(\lambda)\sigma] + \alpha\lambda \\ \varepsilon_3 = \dfrac{1}{E(\lambda)}[(1+\nu(\lambda))\sigma_3 - \nu(\lambda)\sigma] + \alpha\lambda \end{cases} \quad (1.76)$$

The total cubic deformation equals:

$$\theta = \varepsilon_1 + \varepsilon_2 + \varepsilon_3 \quad (1.77)$$

Substituting (1.76) in (1.77), we get a dependence of a total volume deformation θ on volume stress σ and the swelling parameter λ in the form:

$$\theta = \frac{1-2\nu(\lambda)}{E(\lambda)}\sigma + 3\alpha\lambda \quad (1.78)$$

hence:

$$\sigma = \frac{E(\lambda)}{1-2\nu(\lambda)}[\theta - 3\alpha\lambda] \quad (1.79)$$

Solving system (1.76) with respect to σ_{ij} and taking into account (1.79), we get dependence of stresses σ_{ij} on the deformation components ε_{ij}, volume deformation θ and the swelling parameter λ in the form:

$$\sigma_{ij} = a(\lambda)\cdot\theta\cdot\delta_{ij} + 2G(\lambda)\varepsilon_{ij} - \frac{E(\lambda)}{1-2\nu(\lambda)}\alpha\lambda\cdot\delta_{ij} \quad (1.80)$$

Here:

$$a(\lambda)=\frac{E(\lambda)\cdot\nu(\lambda)}{(1+\nu(\lambda))(1-2\nu(\lambda))},\ 2G(\lambda)=\frac{E(\lambda)}{1+\nu(\lambda)} \quad (1.81)$$

are generalized Lame coefficients and sear modulus taken for each value of the parameter λ; δ_{ij} is Kronecker's symbol:

$$\delta_{ij}=\begin{cases}1 & for \quad i=j\\ 0 & for \quad i\neq j\end{cases} \quad (1.82)$$

Knowing the approximation functions E(λ) and ν(λ), that is, the values of experimental data $E(\lambda)|_{\lambda=0}=E_0$, $E(\lambda)|_{\lambda=\lambda_{max}}=E_{min}$, $\nu(\lambda)|_{\lambda=0}=\nu_0$, $\nu(\lambda)|_{\lambda=\lambda_{max}}=\nu_{max}$, we give relation (1.80) the another form. In other words, represent relations (1.80) by the correction functions of swelling $\phi(\lambda)$, $\psi(\lambda)$, $\eta(\lambda)$. By experimental data (Figures 1.6 and 1.7), represent the elasticity modulus $E(\lambda)$ depending on the swelling parameter λ in the form of exponential function, the Poisson ratio $\nu(\lambda)$ in the form of a linear function of the form:

$$E(\lambda)=E_0\left(\frac{E_{min}}{E_0}\right)^{\lambda/\lambda_{max}}=E_0\left(\frac{E_{min}}{E_0}\right)^{\tilde{\lambda}}=E_0 b^{\tilde{\lambda}} \quad (1.83)$$

$$\nu(\lambda)=\nu_0+(\nu_{max}-\nu_0)\frac{\lambda}{\lambda_{max}}=\nu_0+k\tilde{\lambda} \quad (1.84)$$

Here, $k=\nu_{max}-\nu_0$ is a dimensionless parameter, $\tilde{\lambda}=\frac{\lambda}{\lambda_{max}}$ changes in the interval $0\le\tilde{\lambda}\le 1$. In this case, the Lame generalized coefficient (1.81) and the function $\tilde{\eta}(\lambda)=\frac{E(\lambda)}{1-2\nu(\lambda)}\cdot\lambda$ accept the following from:

$$a(\lambda)=a_0\psi(\lambda),\ 2G(\lambda)=2G_0\phi(\lambda),\ \tilde{\eta}(\lambda)=\eta_0\cdot\eta(\lambda) \quad (1.85)$$

Here, $2G_0 = \frac{E_0}{1+\nu_0}$, $a_0 = \frac{E_0\nu_0}{(1+\nu_0)(1-2\nu_0)}$ are Lame coefficients ignoring swelling; $\phi(\lambda)$ and $\psi(\lambda)$ and $\eta(\lambda)$ are the correction coefficients that depend on the swelling parameter λ of the material and equal:

$$\phi(\lambda) = \frac{b^{\tilde{\lambda}}}{1+\frac{k}{1+\nu_0}\cdot\tilde{\lambda}},$$

$$\psi(\lambda) = \frac{b^{\tilde{\lambda}}(1+\frac{k}{\nu_0}\cdot\tilde{\lambda})}{(1+\frac{k}{1+\nu_0}\cdot\tilde{\lambda})(1-\frac{2k}{1-2\nu_0}\cdot\tilde{\lambda})} = \phi(\lambda)\cdot\frac{1+\frac{k}{\nu_0}\cdot\tilde{\lambda}}{1-\frac{2k}{1-2\nu_0}\cdot\tilde{\lambda}},$$

$$\eta_0 = \frac{E_0}{1-2\nu_0},\quad \eta(\lambda) = \frac{b^{\tilde{\lambda}}\cdot\lambda_{max}}{1-\frac{2k}{1-2\nu_0}\cdot\tilde{\lambda}} \tag{1.86}$$

where, $\tilde{\lambda} = \lambda/\lambda_{max}$, $k = \nu_{max} - \nu_0$, $b = E_{min}/E_0$. For $\tilde{\lambda} = 0$, $\phi(0) = 1$, $\psi(0) = 1$, $\eta(0) = \lambda_{max}$. For $\tilde{\lambda} = 1$ the correction factors $\phi(\lambda)$, $\psi(\lambda)$, and $\eta(\lambda)$ will be in the form:

$$\phi(\tilde{\lambda})\Big|_{\tilde{\lambda}=1} = \frac{1}{2G_0}\cdot\frac{E_{min}}{1+\nu_{max}},$$

$$\psi(\tilde{\lambda})\Big|_{\tilde{\lambda}=1} = \frac{1}{a_0}\cdot\frac{E_{min}\nu_{max}}{(1+\nu_{max})(1-2\nu_{max})},$$

$$\eta(\tilde{\lambda})\Big|_{\tilde{\lambda}=1} = \frac{1}{\eta_0} \cdot \frac{E_{\min}\lambda_{\max}}{1-2\nu_{\max}} \tag{1.87}$$

The domains of these correction factors will be in the following form. By changing the swelling parameter $\tilde{\lambda} = \dfrac{\lambda}{\lambda_{\max}}$ in the interval $0 \le \tilde{\lambda} = \dfrac{\lambda}{\lambda_{\max}} \le 1$, the correction factors $\phi(\lambda)$, $\psi(\lambda)$ and $\eta(\lambda)$ will change in the following intervals:

$$1 \ge \phi(\tilde{\lambda}) \ge \frac{1}{2G_0} \cdot \frac{E_{\min}}{1+\nu_{\max}},$$

$$1 \le \psi(\tilde{\lambda}) \le \frac{1}{a_0} \cdot \frac{E_{\min}\nu_{\max}}{(1+\nu_{\max})(1-2\nu_{\max})},$$

$$\lambda_{\max} \ge \eta(\tilde{\lambda}) \ge \frac{1}{\eta_0} \cdot \frac{E_{\min}\lambda_{\max}}{1-2\nu_{\max}} \tag{1.88}.$$

Taking into account formula (1.86), relations (1.80) will be a connection between total stress and deformations for polymer material in the form of Hooke's generalized law with regard to influence of corrosive liquid medium, that were revised by means of correction factors $\phi(\lambda)$, $\psi(\lambda)$, $\eta(\lambda)$, in the form:

$$\sigma_{ij} = a_0 \cdot \psi(\lambda) \cdot \theta \cdot \delta_{ij} + 2G_0\phi(\lambda) \cdot \varepsilon_{ij} - \eta_0\eta(\lambda)\alpha\tilde{\lambda}\delta_{ij} \tag{1.89}$$

From the formula (1.33) and (1.37), express $E(\lambda)$ and $\nu(\lambda)$ by the introduced correction functions $\phi(\lambda)$ and $\psi(\lambda)$, in the form:

$$E(\lambda) = 3G_0\phi(\lambda)\frac{1+\dfrac{2G_0}{3a_0}\dfrac{\phi(\lambda)}{\psi(\lambda)}}{1+\dfrac{G_0}{a_0}\dfrac{\phi(\lambda)}{\psi(\lambda)}},$$

$$\nu(\lambda) = \frac{1}{2[1 + \frac{G_0}{a_0}\frac{\phi(\lambda)}{\psi(\lambda)}]} \tag{1.90}$$

In this case, dependence of volume stress $\sigma = \sigma_{ii}$ (1.79) on volume deformation θ and the swelling parameter l with regard to correction factors (1.90) will take the form:

$$\sigma(\varepsilon,\lambda) = \sigma_{ij} = 3a_0\psi(\lambda)[1 + \frac{2G_0}{3a_0}\frac{\phi(\lambda)}{\psi(\lambda)}](\theta - 3\tilde{\alpha}\tilde{\lambda})$$

where, $\tilde{\alpha} = \alpha\lambda_{\max}$. Thus, Hooke's generalized law for a polymer material with regard to effect of change of its physico–chemical property, in the final form will be:

$$\begin{cases} \sigma_{ij} = a_0 \cdot \psi(\lambda) \cdot \theta \cdot \delta_{ij} + 2G_0\phi(\lambda) \cdot \varepsilon_{ij} - \eta_0\eta(\lambda)\alpha\tilde{\lambda} \cdot \delta_{ij} \\ \sigma(\varepsilon,\lambda) = \sigma_{ij} = 3a_0\psi(\lambda)[1 + \frac{2G_0}{3a_0}\frac{\phi(\lambda)}{\psi(\lambda)}](\theta - 3\tilde{\alpha}\tilde{\lambda}) \end{cases} \tag{1.91}$$

where, the correction factors $\phi(\lambda)$, $\psi(\lambda)$ and $\eta(\lambda)$ change in the following limits:

$$1 \geq \phi(\tilde{\lambda}) \geq \frac{1}{2G_0} \cdot \frac{E_{\min}}{1+\nu_{\max}},$$

$$1 \leq \psi(\tilde{\lambda}) \leq \frac{1}{a_0} \cdot \frac{E_{\min}\nu_{\max}}{(1+\nu_{\max})(1-2\nu_{\max})}$$

$$\lambda_{\max} \geq \eta(\tilde{\lambda}) \geq \frac{1}{\eta_0} \cdot \frac{E_{\min}\lambda_{\max}}{1-2\nu_{\max}}. \tag{1.92}$$

In the applied plan, this means the following. By calculating stresses and forces in two and three-dimensional problems of mechanics of deformable polymer body with regard to influence of corrosive external medium, it is necessary to use relations (1.91) within the change of correction factors $\phi(\lambda)$, $\psi(\lambda)$ and $\eta(\lambda)$ in the form (1.92).

Numerical values of correction factors for maximal value of the parameter $\tilde{\lambda} = \frac{\lambda}{\lambda_{\max}} = 1$ are represented in Table 1.5 for a number of polymer materials.

TABLE 1.5 Numerical values of correction factors $\phi(\lambda)$, $\psi(\lambda)$, $\eta(\lambda)$ for the value of parameter $\tilde{\lambda} = \frac{\lambda}{\lambda_{\max}} = 1$

Brand	PA	PEHD	PELD	PS	PP
α	0.3333	0.3618	0.3452	0.2333	0.3000
$\lambda_{\max}$	0.0280	0.0275	0.0350	0.0200	0.0200
E_0 (MPa)	133	58	34	153.5	78
ν_0	0.41	0.39	0.4	0.33	0.37
$E_{\min}$	101	44	21	135	64
$\nu_{\max}$	0.440	0.415	0.435	0.370	0.440
$2G_0$ (MPa)	94.3262	41.7266	24.2857	115.4135	56.9343
a_0 (MPa)	214.8542	73.9699	48.5714	112.019	81.0219
η_0 (MPa)	738.8889	263.6364	170.0	451.4705	300.0
$\phi(\lambda)\vert_{\tilde{\lambda}=1}$	0.7436	0.7452	0.6026	0.8538	0.7806
$\psi(\lambda)\vert_{\tilde{\lambda}=1}$	1.1970	1.0262	1.0082	1.2518	2.0113
$\eta(\lambda)\vert_{\tilde{\lambda}=1}$	0.0319	0.0270	0.0333	0.0230	0.0356

TABLE 1.5 *(Continued)*

$\gamma_0(\frac{i}{m^3})$	10791	9319.5	9025.2	10300s.5	8829
$\gamma_0(\frac{kgf}{m^3})$	1100	950	920	1050	900

Below, on the basis of Table 1.5 on a specific example we will numerically show quality and quantity influence of corrosive liquid medium, that is, influence of thrust forces on the character of defining relations.

Example 5: Using the numerical data of Table 1.4, write the models of deformation of polymer material of the brand PS with regard to and regardless of the influence of corrosive liquid medium, that is, thrust forces for $\tilde{\lambda}=0$ and $\tilde{\lambda}=1$. For that, by substituting appropriate numerical values of polymer material of the brand PS to the general deformation model of polymer material (1.91), we get the following special deformation models:

- Deformation model of polymer material of the brand PS ignoring the influence of corrosive liquid medium, that is, for $\tilde{\lambda}=0$, will be in the form:

$$\begin{cases}\sigma_{ij}=112.0190\theta\delta_{ij}+115.4135\varepsilon_{ij}\\ \sigma=451.4590\theta\end{cases} \tag{1.93}$$

- Deformation model of polymer material of the brand PS with regard to influence of corrosive liquid medium, that is, for $\tilde{\lambda}=1$, will be:

$$\begin{cases}\sigma_{ij}=140.2254\cdot\theta\cdot\delta_{ij}+98.54\cdot\varepsilon_{ij}-2.4225\cdot\delta_{ij}\\ \sigma=519.1984\cdot\theta-7.2688\end{cases} \tag{1.94}$$

Comparing the appropriate coefficients in the model represented by formulae (1.93) and (1.94) we can note the followings:

- The coefficient $a(\lambda)|_{\tilde{\lambda}=1}$ corresponding to the case of influence of corrosive liquid medium, that is, the case of influence of swelling of polymer material is higher by 25% than the coefficient $a_0|_{\tilde{\lambda}=0}$ that corresponds to the case of regardless of the influence of corrosive liquid medium, that is, the case when there is no influence of swelling of polymer material.

- Shear modulus $2G(\lambda)|_{\bar{\lambda}=1}$ is less by 14.6% than the case of regardless of influence of liquid medium $2G(\lambda)|_{\bar{\lambda}=0}$, that is, the case regardless of the influence of swelling of polymer material.
- Bulk modulus is greater by 15% than the case of regardless of influence of liquid medium $K(\lambda)|_{\bar{\lambda}=1}, K(\lambda)|_{\bar{\lambda}=0}$.
- Therewith, from the action of swelling forces in polymer material there also arises initial stress of quantity 2.4225 MPa .

On the bases of the stated ones we can make the following conclusion. For investigating strength problems of constructions intended to work in corrosive medium, it is necessary to use defining relations (1.91), and for the considered special case, formula (1.94).

1.2.3 ON CHANGEABILITY OF MASS FORCE OF POLYMER MATERIAL WITH REGARD TO CHANGE OF PHYSICO–CHEMICAL PROPERTIES

One of the features of polymer material contacting with corrosive liquid and gassy media is changeability of its mass force. Gassy and liquid media intensively diffuse into internal layers of polymer material, change its chemical composition, cause swelling effect of the material and this reduces to change of the volume and density of polymer material. In this connection, below we'll define the quantity of the arising thrust mass force of polymer material with regard to change of its physico–chemical properties, and estimate the quality and quantity sides of this effect [5, 9, 12].

For that consider a polymer material with initial volume V_0 and specific weight $\gamma_0 = \rho_0 g$. Here ρ_0 is the density of the material. In this case, the sample's weight will equal (Aliyev, 1995, 2012, 2012):

$$Q_0 = \gamma_0 V_0 \tag{1.95}$$

Let a sample stay in the corrosive medium t time. Under the action of diffusion of corrosive liquid media, the density and volume of polymer material will change. In this case, the material's weight for the value of time t =t* will take the form:

$$Q(\lambda) = (\gamma_0 + \Delta\gamma)(V_0 + \Delta V) = \gamma_0 V_0 (1 + \frac{\Delta\gamma}{\gamma_0})(1 + \frac{\Delta V}{V_0}) \tag{1.96}$$

Here, $\Delta\gamma = \gamma(\lambda) - \gamma_0$, $\Delta V = V(\lambda) - V_0$. Take into account formula (1.51) and (1.53) that are equal to:

$$\varepsilon_V = \frac{\Delta V}{V_0} = 3\alpha\lambda \tag{1.97}$$

$$\frac{\gamma(\lambda)}{\gamma_0} = \frac{1+\lambda}{1+3\alpha\lambda} \tag{1.98}$$

From (1.98) we have:

$$\gamma(\lambda) = \gamma_0 \frac{1+\lambda}{1+3\alpha\lambda} \tag{1.99}$$

Then the absolute change of specific weight $\Delta\gamma$ allowing for (1.52) and (1.53) will be in the form:

$$\Delta\gamma = \gamma_0 \frac{1-3\alpha}{1+3\alpha\lambda}\lambda \tag{1.100}$$

Substitute (1.52) and (1.100) in (1.96) and get:

$$Q(\lambda) = (1+\lambda)Q_0 \tag{1.101}$$

Define the arising quantity of the changed mass force of polymer material $\Delta Q(\lambda)$ define from (1.95) and (1.101) in the form:

$$R(\lambda) = \Delta Q(\lambda) = \lambda Q_0 \text{ for } 0 \le \lambda \le \lambda^* \tag{1.102}$$

Formula (1.102) is the quantity of thrust mass force that arises as a result of swelling of polymer material situated in the corrosive liquid medium. For $\lambda = 0$, the thrust mass force $R(\lambda) = \lambda Q_0$ equals zero. For λ the thrust mass force equal its own maximal value λ*, accepts its own maximal value:

$$R(\lambda) = Q_0 \lambda^* \tag{1.103}$$

Substituting (1.95) in (1.102), the thrust mass force $R(\lambda)$ may be represented in the form:

$$R(\lambda) = \gamma_0 V_0 \lambda \tag{1.104}$$

A force vector of the form (1.102) or (1.104) acting on the volume V_0 having the mass $m_0 = \gamma_0 V_0$ is a vector of thrust mass force at the point to which the volume V_0 concentrates. Thus, knowing the initial weight of the construction $Q_0 = \gamma_0 V_0$ and the swelling parameter λ, the quantity of thrust mass force will be defined by formula (1.104). Hence it follows that the arising thrust mass force $R(\lambda)$ is proportional to the swelling parameter λ of the material. Dimension of this force will be *(kqf)*.

For specific from of the structural element define the kind of thrust mass force and give appropriate quality and quantity analysis. As in the subsequent chapters we'll investigate the problems of mechanics of sandwich pipes therefore in place of structural element we will take one layer and sandwich pipe.

Consider a one-layer pipe made of polymer material of internal radius r_1 and external radius r_2 (Figure 1.9). Let the pipe's material have the specific weight γ_0 and swelling parameter λ .In this case, the arising thrust mass force $R(\lambda)$ per all section and unit length will equal (1.104):

$$R(\lambda) = \gamma_0 V_0 \lambda = \pi (r_2^2 - r_1^2) \cdot \gamma_0 \lambda \cdot 1 \tag{1.105}$$

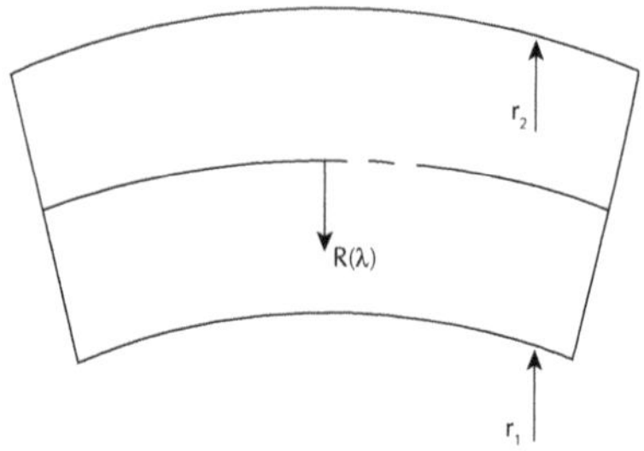

FIGURE 1.9 Dependence of the thrust mass force in one-layer pipe made of polymer material

Then we define the thrust mass force per unit length of the arch of median cross section line from the relation:

$$\begin{cases} R(\lambda) \to 2\pi r_{averag.} = 2\pi \frac{r_2 + r_1}{2} = \pi(r_2 + r_1) \\ R^0(\lambda) \to r_{averag.} \cdot d\theta = \frac{r_2 + r_1}{2} \cdot d\theta \end{cases}$$

when:

$$R^0(\lambda) = \frac{r_{averag.} \cdot d\theta}{\pi(r_2 + r_1)} \cdot R(\lambda) \qquad (1.106)$$

Here, $r_{averag.} \cdot d\theta = \frac{r_2 + r_1}{2} d\theta$ is a unit length of the arch of the median cross section line of the pipe. Thus, the thrust mass force per unit length of the arch of median cross section line in the final form will be as:

$$R^0(\lambda) = \gamma_0 \lambda (r_2 - r_1) \cdot (r_{averag.} d\theta) \qquad (1.107)$$

For a unit length of the arch $r_{averag.} \cdot d\theta$ accept 1cm, that is, $\mathrm{r_{averag.}} \cdot \mathrm{d}\theta = 1\mathrm{cm} = 0.01\mathrm{m}$. Then the thrust mass force per unit length of the arch of the median cross section line of the pipe (1.108) will equal:

$$\mathrm{R}^0(\lambda) = 0.01(\mathrm{r}_2 - \mathrm{r}_1) \cdot \gamma_0 \lambda \qquad (1.108)$$

Numerical calculation for a one-layer pipe made of polymer material of thickness $3\,\text{sm}$ is given in Table 1.6.

TABLE 1.6 Quantity of the thrust mass force in one-layer cylinder made of polymer material of thickness 3cm with regard to effect of change of physico–chemical property

Brand	γ (N/m^3)	λ	r_1 (m)	r_2 (m)	$R(\lambda)$ (N)	$R^0(\lambda)$ (N)
PS	10300.5	0.02	0.1	0.13	4.4704	0.0618
PELD	9025.2	0.035	0.1	0.13	6.8546	0.0948
PEHD	9319.5	0.0275	0.1	0.13	5.5614	0.0769
PA	10791	0.028	0.1	0.13	6.5566	0.0906
PP	8829	0.02	0.1	0.13	3.8318	0.0530

It is seen from the table that in the cylinder made of polymer material of thickness 3cm there arises the following thrust mass force per 1cm of the length of the arch of median line $R^0(\lambda)$: in the brand PS—0.0618N (6.3 gf); in the brand PELD—0.0948N (9.7 gf); in the brand PEHD—0.0769N (7.8 gf); in the brand PA—0.091N (9.3 gf); in the brand PP—0.053N (5.4 gf).

Now define an expression for thrust mass force in cylindrical layers of *n*-layer pipe. For that apply (1.104) to the layers of (n–1)th sandwich pipe. Write (1.104) for the $(n-1)$th and nth layer of the pipe (Figure 1.10).

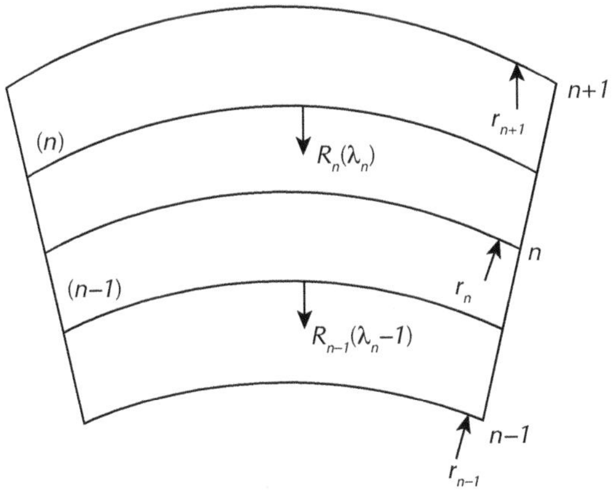

FIGURE 1.10 Dependence of the thrust mass force in two-layer pipe made of polymer material

They will equal:

$$R_n(\lambda_n) = \gamma_{0n} V_{on} \lambda_n \tag{1.109}$$

$$R_{n-1}(\lambda_{n-1}) = \gamma_{0n-1} V_{on-1} \lambda_{n-1} \tag{1.110}$$

Here V_{on} and V_{on-1} are the volumes of the *n*th and (*n* – 1)th layer of a unit length pipe. Then the thrust mass force per annular areas of the *n*th and (*n* – 1)th layers of a unit length pipe $R_n(\lambda_n)$ and $R_{n-1}(\lambda_{n-1})$ will be in the form:

$$R_n(\lambda_n) = \pi r_n^2 (\frac{r_{n+1}^2}{r_n^2} - 1) \cdot \gamma_{0n} \lambda_n \tag{1.111}$$

$$R_{n-1}(\lambda_{n-1}) = \pi r_n^2 (1 - \frac{r_{n-1}^2}{r_n^2}) \cdot \gamma_{0n-1} \lambda_{n-1} \tag{1.112}$$

Then the difference of the mass thrust force $\Delta R_n(\lambda_n) = R_n^-(\lambda_n) - R_{n-1}^+(\lambda_{n-1})$ acting on the *n*th boundary will be represented in the following form (Figure 1.10):

$$\Delta R_n(\lambda_n) = R_n^-(\lambda_n) - R_{n-1}^+(\lambda_n) = \pi r_n^2 [\gamma_{0n} \lambda_n \cdot \\ \cdot (\frac{r_{n+1}^2}{r_n^2} - 1) - \gamma_{0n-1} \lambda_{n-1} \cdot (1 - \frac{r_{n-1}^2}{r_n^2}] \tag{1.113}$$

Now define the thrust mass force per unit length of the arch $R_n^0(\lambda_n)$. The thrust mass force per annular domain of the *n*th and (*n* – 1)th layers will be represented by formulae (1.111) and (1.113). For determining the thrust mass force per unit length of the arch $R_n^0(\lambda_n)$, we behave as follows.

$$\begin{cases} R_n(\lambda_n) \rightarrow 2\pi r_{n,averag.} \\ R_n^0(\lambda_n) \rightarrow dS_{n,averag.} \end{cases} \rightarrow whence$$

$$R_n^0(\lambda_n) = R_n(\lambda_n) \cdot \frac{dS_{n,averag.}}{2\pi r_{n,averag.}} \tag{1.114}$$

where, $dS_{n,averag.} = r_{n,averag.} \cdot d\theta = \frac{r_{n+1} + r_n}{2} d\theta$ is a unit length of the arch of the media line of the *n*th layer. Then we can represent (1.114) in the form:

$$R_n^0(\lambda_n) = \frac{1}{2\pi} R_n(\lambda_n) \cdot d\theta \tag{1.115}$$

Substituting (1.111) in (1.115), we get:

$$R_n^0(\lambda_n) = \frac{1}{2} r_n \left(\frac{r_{n+1}^2}{r_n^2} - 1\right) \cdot \gamma_{0n} \lambda_n \cdot (r_n d\theta) \tag{1.116}$$

Similarly, we write for the $(n-1)$th layer as well:

$$\begin{cases} R_{n-1}(\lambda_{n-1}) \to 2\pi r_{n-1,averag.} \\ R_{n-1}^0(\lambda_{n-1}) \to dS_{n-1,averag.} \end{cases}$$

hence:

$$R_{n-1}^0(\lambda_{n-1}) = R_{n-1}(\lambda_n) \cdot \frac{dS_{n-1,averag.}}{2\pi r_{n-1,averag.}} \tag{1.117}$$

where, $dS_{n-1,averag.} = r_{n-1,averag.} \cdot d\theta = \frac{r_n + r_{n-1}}{2} d\theta$ is a unit length of the arch of the median line of the $(n-1)$th layer. Then we represent (1.116) in the form:

$$R_{n-1}^0(\lambda_{n-1}) = \frac{1}{2\pi} R_{n-1}(\lambda_{n-1}) \cdot d\theta \tag{1.118}$$

Substituting (1.112) in (1.118), we get:

$$R_{n-1}^{0}(\lambda_{n-1}) = \frac{1}{2} r_n \left(1 - \frac{r_{n-1}^2}{r_n^2}\right) \cdot \gamma_{0n-1} \lambda_{n-1} \cdot (r_n d\theta) \qquad (1.119)$$

So, we obtained the following expressions of a unit thrust mass force per unit length of the arch in the *n*th and (*n* – 1)th layers, in the form:

$$\begin{cases} R_n^0(\lambda_n) = \frac{1}{2} r_n \left(\frac{r_{n+1}^2}{r_n^2} - 1\right) \cdot \gamma_{0n} \lambda_n \cdot (r_n \cdot d\theta) \\ R_{n-1}^0(\lambda_{n-1}) = \frac{1}{2} r_n \left(1 - \frac{r_{n-1}^2}{r_n^2}\right) \cdot \gamma_{0n-1} \lambda_{n-1} \cdot (r_n \cdot d\theta) \end{cases} \qquad (1.120)$$

Here $r_n \cdot d\theta$ is a unit length of the arch per the *n*th boundary surface. The convenience of such a representation is that by its help we can show the influence of thrust mass force of the *n*th and (*n* – 1)th layers with respect to boundary surface between them. In this case, the difference of unit thrust mass forces $\Delta R_n^0(\lambda_n) = R_n^{0,-}(\lambda_n) - R_{n-1}^{0,+}(\lambda_{n-1})$ will equal:

$$\Delta R_n^0(\lambda_n) = \frac{1}{2} r_n \left[\left(\frac{r_{n+1}^2}{r_n^2} - 1\right) \cdot \gamma_{0n} \lambda_n - \right.$$
$$\left. -\left(1 - \frac{r_{n-1}^2}{r_n^2}\right) \cdot \gamma_{0n-1} \lambda_{n-1}\right] \cdot (r_n d\theta) \qquad (1.121)$$

For a unit length of the arch $(r_n \cdot d\theta)$ accept 1cm, that is $r_n.d\theta$ = 1cm = 0.01m. Then the difference of thrust mass force per the *n*th boundary surface will be:

$$\Delta R_n^0(\lambda_n) = \frac{1}{200} r_n \left[\left(\frac{r_{n+1}^2}{r_n^2} - 1\right) \cdot \gamma_{0n} \lambda_n - \left(1 - \frac{r_{n-1}^2}{r_n^2}\right) \cdot \gamma_{0n-1} \lambda_{n-1}\right] \qquad (1.122)$$

The numerical values of arising thrust mass forces under the contact of polymer material with corrosive oily liquid, calculated by (1.118) are given in Table 1.7 for a class of polymer material. So, in elementary element of cylindrical interlayer made of polymer material of thickness 3cm, there arises the following thrust mass force per 1cm of length of the arch $R_n^0(\lambda_n)$: in the brand PS—0.0711N (7.2 gf); in the brand PELD—0.109N (1.1gf); in the brand PEHD—0.0884N (9 gf); in the brand PA—0.1042N (10.6gf); in the brand PP—0.0609N (6.2 gf).

TABLE 1.7 Quantity of thrust mass force in cylindrical interlayer made of polymer material of thickness 3cm with regard to the factor of change of its physico–chemical properties

Brand	γ_{0n} (N/m^3)	λ_n	r_n (m)	r_{n+1} (m)	$R_n(\lambda_n)$ (N)	$R_n^0(\lambda_n)$ (N)
PS	10300.5	0.02	0.1	0.13	4.4634	0.0711
PELD	9025.2	0.035	0.1	0.13	6.8439	0.1090
PEHD	9319.5	0.0275	0.1	0.13	5.5527	0.0884
PA	10791	0.028	0.1	0.13	6.5463	0.1042
PP	8829	0.02	0.1	0.13	3.8258	0.0609

We also show the numerical values of the difference of thrust mass force that arises in contact problems of sandwich pipes made of polymer materials. For that consider a two layer pipe whose internal layer was made from polymer material of the brand PELD and external layer from the material PS with the following geometric and physico–chemical properties:

$$r_1 = 0.1\text{m},\ r_2 = 0.13\text{m},\ r_3 = 0.15\text{m};$$

$$\gamma_{01} = 9025.2\frac{N}{m^3},\ \lambda_1 = 0.0288;\ \gamma_{02} = 10300.5\frac{N}{m^3},$$

$$\lambda_2 = 0.02; \tag{1.123}$$

Calculating by formula (1.122) the difference of thrust mass force arising on the interface of two layer pipe, per unit length of 1cm, will equal 0.025N (2.5 gf). In other words, because of different type influences of physico–chemical change of materials in the layers, on the interface of two layers made of polymer material the contact stress $\Delta\sigma_n = \sigma_n^+ - \sigma_n^-$ will have a break by the quantity 0.025N (2.5 gf).

1.2.4 ELASTICO–PLASTIC DEFORMATION MODEL OF POLYMER MATERIAL WITH REGARD TO CHANGE OF PHYSICO–CHEMICAL PROPERTIES

A large class of polymer materials behaves as a non linear-elastic body. In this section, we shall investigate objective laws of non linear physico–chemical behavior of polymer material with regard to influence of corrosive liquid medium effect, that is, the swelling effect of the material (Aliyev, 1984, 1987, 1995, 2012, 2012).

Let a polymer material in the corrosive medium be subjected to the law of deformation of deformation type with linear strengthening (Ilyushin, 1948; Ilyushin and Lenskiy, 1959). For each value of the swelling parameter λ_n, the experimental curves of mechanical changes of the material's properties will be described by the appropriate diagrams represented on the Figure 1.11.

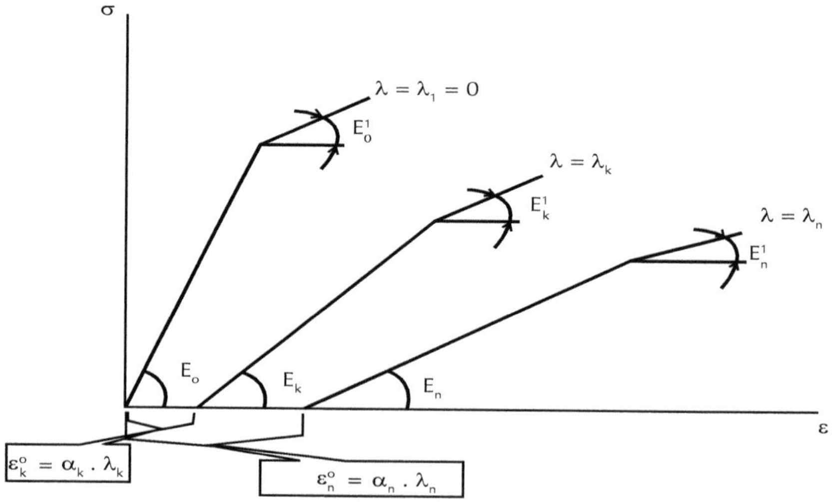

FIGURE 1.11 Diagram of $\sigma = f(\varepsilon)$ for strengthening material under different values of swelling parameter λ

According to the figure, common longitudinal deformation of the sample e will be composed of deformation of only swelling $\varepsilon_0 = \alpha\lambda$

and deformation arising only from the mechanically applied load $\varepsilon_1(P) = \varepsilon - \varepsilon_0$, in the form (Figure 1.12):

$$\varepsilon = \varepsilon_0(\lambda) + \varepsilon_1(P) \tag{1.124}$$

Here, $\varepsilon_0 = \alpha\lambda$ is the deformation only from the swelling of polymer material, $\varepsilon_1 = \varepsilon - \alpha\lambda$ is the elastic-plastic deformation of the sample arising only from the mechanically applied load, σ_s and $\varepsilon_s = \varepsilon_1 - \varepsilon_0$ are the coordinates of the point K that corresponds to the yield point of the material. Here, the quantity of deformation ε_s is the yield point of the material arising only from the action of mechanically applied load.

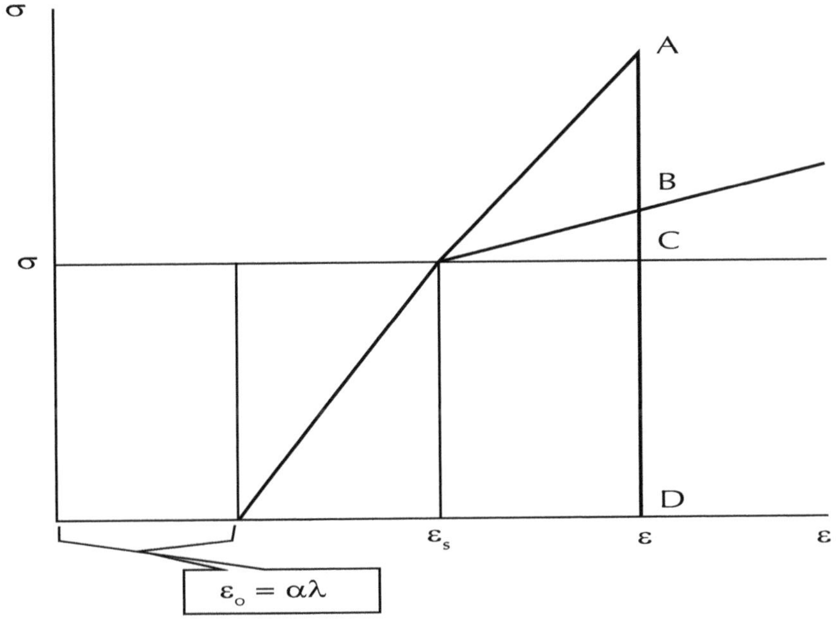

FIGURE 1.12 The graph of elastico-plastic material $\sigma = (\varepsilon)$ with regard to change of physico–chemical property

From Figure 1.12 we have:

$$BD = AD - AC + BC \tag{1.125}$$

where, $AD = E_0(\lambda)\cdot MD$, $AC = E_0(\lambda)\cdot KC$, $BC = E'(\lambda)\cdot KC$, $\varepsilon_0 = \alpha\lambda$, $\varepsilon_s = \varepsilon_1 - \alpha\lambda$, $MD = \varepsilon - \alpha\lambda$, $KC = \varepsilon - \varepsilon_1$ (1.126)

Allowing for (1.126), formula (1.125) takes the form:

$$\sigma = E_0(\lambda)[1-(1-\frac{E'(\lambda)}{E_0(\lambda)})\frac{1-\frac{\varepsilon_1}{\varepsilon}}{1-\frac{\alpha\lambda}{\varepsilon}}]\cdot(\varepsilon-\alpha\lambda) \quad (1.127)$$

Expanding the function $\frac{1}{1-\frac{\alpha\lambda}{\varepsilon}}$ in series and taking into account $\frac{\alpha\lambda}{\varepsilon}<1$, we get:

$$\frac{1}{1-\frac{\alpha\lambda}{\varepsilon}} = 1+\frac{\alpha\lambda}{\varepsilon} \quad (1.128)$$

Substituting (1.128) in (1.127) and taking into account $\frac{\varepsilon_s}{\varepsilon}\cdot\frac{\alpha\lambda}{\varepsilon} << 1$ we get:

$$\sigma = E_0(\lambda)[1-(1-\frac{E'(\lambda)}{E_0(\lambda)})(1-\frac{\varepsilon_1-\alpha\lambda}{\varepsilon})]\cdot(\varepsilon-\alpha\lambda)$$

Taking into account $\varepsilon_s = \varepsilon_1 - \alpha\lambda$ we get:

$$\sigma = E_0(\lambda)[1-(1-\frac{E'(\lambda)}{E_0(\lambda)})(1-\frac{\varepsilon_s}{\varepsilon})]\cdot(\varepsilon-\alpha\lambda) \quad (1.129)$$

Introduce the denotation:

$$\omega(\varepsilon,\lambda) = (1-\frac{E'(\lambda)}{E_0(\lambda)})(1-\frac{\varepsilon_s}{\varepsilon}) \quad (1.130)$$

Then, allowing for (1.130), represent (1.129) in the following final form:

$$\sigma(\varepsilon,\lambda) = E_0(\lambda)[1-\omega(\varepsilon,\lambda)]\cdot(\varepsilon-\alpha\lambda) \tag{1.131}$$

Here $\omega(\lambda,\varepsilon)$ is the polymer material strengthening function with regard to the effect of change of physico–chemical characteristics of the material. Hence it is seen that the polymer material strengthening function $\omega(\lambda,\varepsilon)$, by physico–chemical change of polymer material is always less than the quantity $\omega(0,\varepsilon)$, that is, $\omega(\lambda,\varepsilon)\le\omega(0,\varepsilon)$. For $\lambda=0$, model (1.127) coincides with A. Ilyushin's strengthening model (Ilyushin, 1948, Ilyushin and Lenskiy, 1959).

1.2.5 EXPERIMENTAL DETERMINATION OF TEMPERATURE INFLUENCE OF VELOCITY OF CHANGE OF PHYSICO–CHEMICAL PROPERTIES OF POLYMER MATERIALS SITUATED IN CORROSIVE LIQUID MEDIUM

Above we have shown the influence of corrosive liquid, in particular of oil on physico–chemical and physico–mechanical properties of polymer materials under normal temperature. On this basis, appropriate physico–mechanical models of deformation of polymer material with regard to effect of diffusion of liquid media were suggested.

In this section, we will represent experimental investigations on establishing the influence of corrosive liquid medium temperature on the swelling process and change of physico–chemical properties of polymer materials (Aliyev, 1995, 2012, 2012; Aliyev and Gabibov, 1994).

Given the samples made of polymer materials placed in corrosive liquid that have different temperatures. Let the corrosive medium temperature have the values $T_1, T_2, T_3,\ldots, T_n$. Depending on time and temperature, the samples placed in the corrosive medium will swell. And the swelling velocity of polymer material will display essential dependence on the quantity of temperature. Therefore for different fixed values of temperature $T_1, T_2, T_3,\ldots, T_n$, we construct the following diagrams:

$$\lambda(t,T_k)=\frac{Q(t,T_k)-Q_0}{Q_0} \quad (1.132).$$

In place experimental samples the following polymer materials were used: polyethylene of lower density (PELD), polypropylene (PP), polyamide (KK) and polystyrene (PS). The oil of dynamic velocity $\mu=32.41\times10^{-3}\,Pa\times s$ with 0.277% asphaltene acids (Figure 1.13–1.16, Table. 1.8) was taken in place of corrosive liquid medium. It is seen from the figures that changes of physico–chemical properties of polymers essentially depend on the level of medium's temperature. In other words, by increasing the medium's temperature, the process of diffusion of corrosive liquid medium into polymer material essentially accelerates and a polymer swells significantly. So, polymer samples maintained in aggressive medium 20hrs, under the mediums temperature 60°C, the value of the swelling parameter λ is three times higher than of normal one. Secondly, it was established that swelling velocities of polymer materials maintained in corrosive medium during 2hrs, are non linearly dependent on time. After that the swelling velocity practically for all temperatures remains constant, that is, for $\lambda>\lambda_s$ velocity α' will be in the form:

$$\alpha'(\lambda)=\frac{\lambda(t)-\lambda_s}{t-t_s}=const \quad (1.133)$$

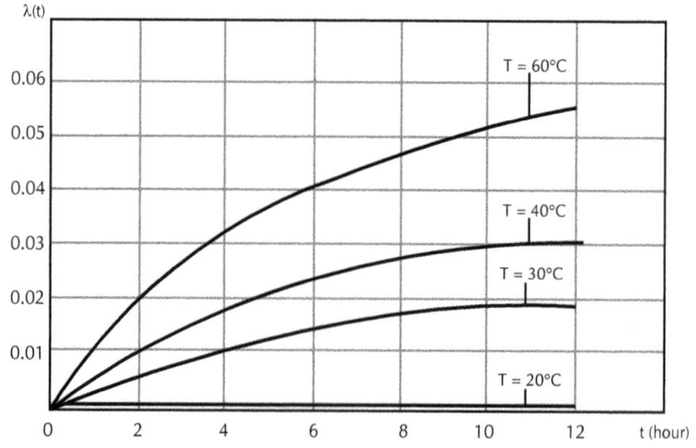

FIGURE 1.13 Dependence of swelling function $\lambda(t)$ on time under different temperatures for polyamide 6 (K)

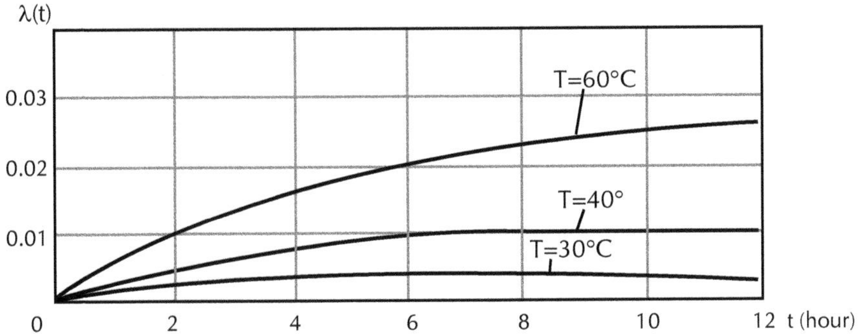

FIGURE 1.14 Dependence of swelling function $\lambda(t)$ on time under different temperatures for a polyethylene of high density (PEHD)

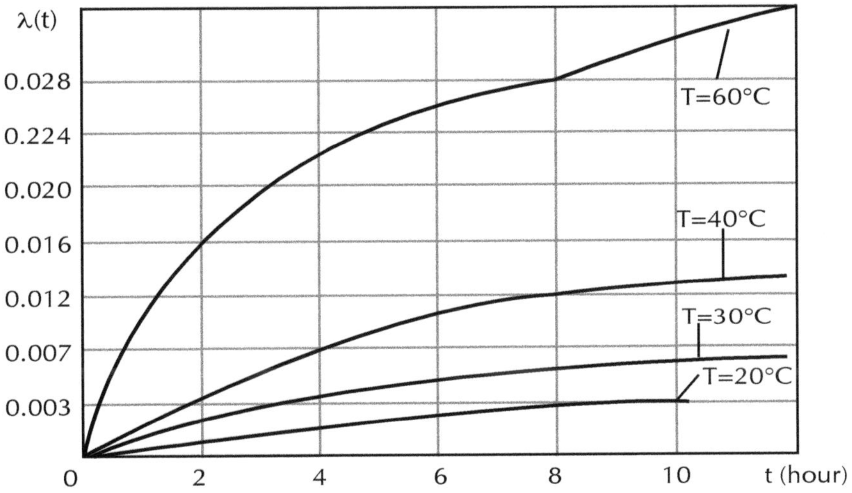

FIGURE1.15 Dependence of swelling function $\lambda(t)$ on time under different temperatures for polystyrene (PS)

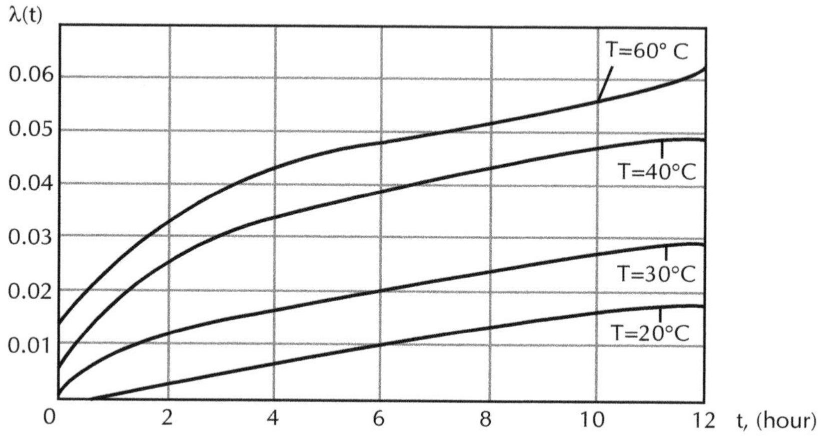

FIGURE 1.16 Dependence of swelling function $\lambda(t)$ on time under different temperatures for PP

TABLE 1.8 Dependences of the swelling function $\lambda(t)$ on time under different temperatures T_n for a polymer material placed in oily medium

T (°C)	t (hr)	$Q(\lambda)$ (gf)	$\lambda(t) = \frac{Q(\lambda)-Q_0}{Q_0}$	$\alpha' = \frac{\lambda(t)-\lambda_s}{\lambda_s}$
1	2	3	4	5
			Polyamide 6 (K)	
	0	0.3290	0	
	2	0.3299	0.0030	
20	4	0.3305	0.0045	2.5×10^{-4}
	8	0.3306	0.0050	
	12	0.3471	0.0055	
	0	0.3273	0	
	2	0.3289	0.0075	
30	4	0.3309	0.0110	6.5×10^{-4}
	8	0.3460	0.0132	
	12	0.3620	0.0145	

TABLE 1.8 *(Continued)*

	0	0.3282	0	
	2	0.3328	0.0140	
40	4	0.3349	0.0205	11.6×10^{-4}
	8	0.3365	0.0265	
	12	0.3379	0.0980	
	0	0.3294	0	
	2	0.3375	0.0240	
60	4	0.3418	0.0378	22.5×10^{-4}
	8	0.3461	0.0477	
	12	0.3463	0.0512	
1	**2**	**3**	**4**	**5**
		PEHD		
	0	0.2725	0	
	2	0.2725	0.0010	
20	4	0.2728	0.0013	
	8	0.2729	0.0016	
	12	0.2730	0.0020	
	0	0.2725	0	
	2	0.2733	0.0031	
30	4	0.2735	0.0038	
	8	0.2738	0.0048	
	12	0.2739	0.0055	
	0	0.2730	0	
	2	0.2743	0.0049	
40	4	0.2749	0.0072	
	8	0.2758	0.0084	
	12	0.2788	0.0105	
	0	0.2730	0	
	2	0.2758	0.0105	
60	4	0.2770	0.0148	
	8	0.2781	0.0188	
	12	0.2794	0.0234	

TABLE 1.8 *(Continued)*

		PS		
	0	0.3050	0	
	2	0.3055	0.0016	
20	4	0.3056	0.0020	2×10^{-4}
	8	0.3058	0.0026	
	12	0.3061	0.0036	
	0	0.3050	0	
	2	0.3061	0.0035	
30	4	0.3064	0.0045	3.9×10^{-4}
	8	0.3069	0.0062	
	12	0.3073	0.0074	

1	**2**	**3**	**4**	**5**
	0	0.3050	0	
	2	0.3069	0.0059	
40	4	0.3078	0.0093	6.8×10^{-4}
	8	0.3086	0.0118	
	12	0.3092	0.0136	
	0	0.3050	0	
	2	0.3100	0.0164	
60	4	0.3112	0.0202	15×10^{-4}
	8	0.3129	0.0259	
	12	0.3148	0.0322	
		PP		
	0	0.2605	0	
	2	0.2607	0.0008	
20	4	0.2607	0.0009	0.475×10^{-4}
	8	0.2608	0.0011	
	12	0.2608	0.0013	

TABLE 1.8 *(Continued)*

	0	0.2610	0	
	2	0.2613	0.0013	
30	4	0.2614	0.0016	1.5×10^{-4}
	8	0.2616	0.0022	
	12	0.2617	0.0028	
	0	0.2610	0	
	2	0.2617	0.0028	
40	4	0.2619	0.0033	1.9×10^{-4}
	8	0.2620	0.0040	
	12	0.2623	0.0048	
	0	0.2610	0	
	2	0.2619	0.0036	
60	4	0.2621	0.0042	2.5×10^{-4}
	8	0.2623	0.0051	
	12	0.2626	0.0061	

1.3 LONGITUDINAL STABILITY OF A STRIP UNDER SWELLING FORCES

It was shown above that the corrosive liquid media cause contractive stresses in structural elements. In this connection, for providing reliable service of structures made of polymer and composite materials maintained in corrosive media, calculated by strength theory, it is necessary to impose additional condition on geometrical characteristics of these constructions that could provide their reliable service for stability. Below we suggest a variant of stability criterion for a strip situated only under the action of swelling forces (Aliyev, 2012, 2012).

Let a strip of length l, width b, thickness h hingely fastened by the end-faces be situated in the corrosive liquid medium. Therewith, there is no longitudinal displacement of end-faces (Figure 1.17). Longitudinal-contractive stress in the strip arises only under the action of swelling effect (Aliyev, 2012).

Define critical compression stress that causes loss of stability of the strip, and also define mathematical dependence of ratio of critical length $\frac{\ell_{cr}}{h}$ to the width on the swelling parameter λ.

The bending equation for a strip for each value of the swelling parameter λ will have the form:

$$E(\lambda)J(\lambda)\frac{d^2y}{dx^2} = M(\lambda) \tag{1.134}$$

Here $J(\lambda) = \frac{b(\lambda)h^3(\lambda)}{12}$ is the moment of inertia of the strips cross section and depends on the swelling parameter λ, $y = y(x)$ is the deflection function of the strip. The dependence of the bending moment will be in the form:

$$M(\lambda) = -P_{contr.}(\lambda)y = -\sigma_{contr.}F(\lambda)\cdot y \tag{1.135}$$

Substituting (1.135) into (1.134), the differential bending equation will take the form:

$$\frac{d^2y}{dx^2} + k^2y = 0,\ k^2 = -\frac{12\sigma(\lambda)}{E(\lambda)h^2} \tag{1.136}$$

The solution of equation (1.136) in the case of hinge support, that is, for $y|_{x=0} = y|_{x=l} = 0$ will be:

$$y = \sin\frac{n\pi}{\ell}x \tag{1.137}$$

Substituting (1.137) in equation (1.136), find the quantity of the contractive stress $\sigma_{contr.}$ in the form:

$$\sigma(\lambda) = \frac{\pi^2 E(\lambda)h^2(\lambda)}{12\ell^2(\lambda)}n^2 \tag{1.138}$$

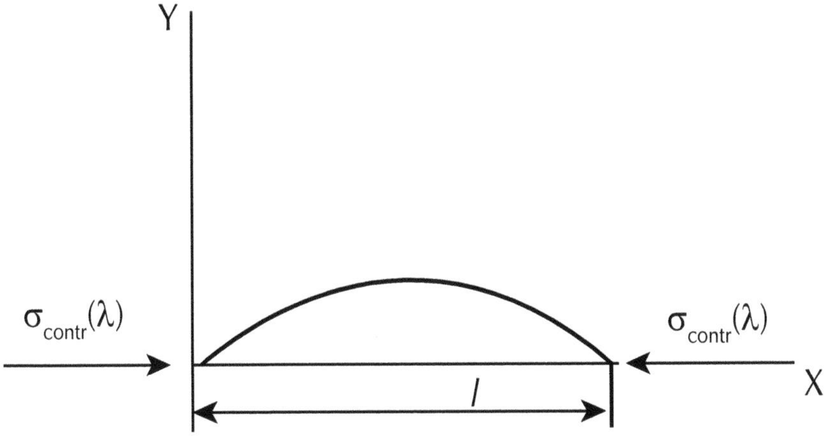

FIGURE 1.17 Buckling of polymer strip under the action of thrust forces arising from swelling effect

For $n = 1$, it holds strip's bending on one half-wave. In this case the value of the compression stress $\sigma(\lambda)$ takes its critical value $\sigma_{cr.}(\lambda)$, causing the loss of stability of the strip, in the form:

$$\sigma(\lambda) = \frac{\pi^2 E(\lambda) h^2(\lambda)}{12\ell^2(\lambda)} n^2 \tag{1.139}$$

By holding geometrical characteristics of the strip $\ell(\lambda)$, $h(\lambda)$ and elasticity modulus $E(\lambda)$ for each value of λ or its limit value λ^*, by formula (1.139) we define the critical value of contractive stress.

Mechanical peculiarity of constructions made of polymer materials operating in corrosive liquid media is that being situated in the corrosive medium, it may be strong but unstable. In other words, if a construction is not under the action of corrosive medium, the construction is always stable and strong. Therefore, such a construction is calculated according to strength criterion, because the construction's serviceability under these external conditions, for the values of stress will be much less than the values of its ultimate strength. Therefore, in this class of problems, it is a matter of principle to find the dependence of the ratio of length to thickness with regard to the swelling parameter. This necessary condition could provide stability of the construction depending on the swelling parameter $\ell_{cr.}/h = f_{cr.}(\lambda)$. In this connection, assuming $\sigma_{cr.}(\lambda)$ to be known from (1.139), we find a condition on the ratio $\frac{\ell}{h}$, in the form:

$$\ell_{cr.}/h = \frac{\pi}{2\sqrt{3}}\sqrt{\frac{E(\lambda)}{\sigma_{cr.}(\lambda)}} \tag{1.140}$$

Using the experimental dependence $E(\lambda) \sim \lambda$, and also calculating $\sigma(\lambda)$ from formula (1.64), condition (1.140) allows to establish the dependence of critical length of the sample $\ell_{cr.}$ on the thickness h of the strip.

It should be noted that the limit case $\lambda = \lambda_{max}$, $E(\lambda_{max}) = E_{min}$, $\sigma(\lambda_{max}) = \sigma_{max}$ allows to determine the critical length $\ell_{cr.}(\lambda_{max})$ that provides stability of the strip for any value of $0 \le \lambda \le \lambda_{max}$ and equals:

$$\ell_{cr.}/h < \frac{\pi}{2\sqrt{3}}\sqrt{\frac{E_{min}}{\sigma_{max}}} \tag{1.141}$$

Substituting (1.64) into (1.140) and (1.141), write the ratio $\ell_{cr.}/h$ depending on the swelling factor of the strip α in the form:

$$\ell_{cr.}/h < \frac{\pi}{2\sqrt{3}}\sqrt{\frac{1}{\alpha\lambda}} \tag{1.142}$$

or

$$\ell_{cr.}/h < \frac{\pi}{2\sqrt{3}}\sqrt{\frac{1}{\alpha\lambda_{max}}} \tag{1.143}$$

The peculiarity of formulae (1.141) and (1.142) is that by knowing physico–chemical changes of the material, it is possible to predict mechanical loss of stability of a strip.

Taking the case of compressed bar with hingley-supported ends in place of the main case, determine the influence of types of fastening of ends to the loss of stability of a strip. Introducing the Yasin quantity (Aliyev, 1995, 1998, 2012, 2012; Aliyev and Gabibov, 1994) of the reduced length (μl), the expression for critical length independent of the fastening type will be of the form:

$$\ell_{cr.}/h < \frac{\pi}{2\mu\sqrt{3}}\sqrt{\frac{E(\lambda)}{\sigma(\lambda)}} \tag{1.144}$$

or

$$\ell_{cr.}/h < \frac{\pi}{2\mu\sqrt{3}}\sqrt{\frac{1}{\alpha\lambda_{max}}} \tag{1.145}$$

TABLE 1.9 Numerical values of critical lengths of a strip for polymer material under the action of oily corrosive medium

Material's brand	E (l) (MPa)	s (l) (MPa)	Hinge ends $l_{cr} = \frac{\pi h}{2\sqrt{3}}\sqrt{\frac{E(\lambda)}{\sigma(\lambda)}}$ (mm)	l_{cr}/h	Built-in ends $l_{cr} = \frac{\pi h}{\sqrt{3}}\sqrt{\frac{E(\lambda)}{\sigma(\lambda)}}$ (mm)	l_{cr}/h
PA	101.5	2.5578	11.34	5.67	22.68	11.34
PALD	20.3	0.4223	12.92	6.46	25.84	12.92
PAHD	44.0	0.6	15.54	7.77	31.08	15.54
PS	125	1	20.29	10.14	40.58	20.29
PP	66.25	0.4041	23.34	11.67	46.68	23.34

For hinge ends (main case) $\mu = 1$; for one free, another built-in $\mu = 2$; for both built-in $\mu = 1/2$ (Table. 1.8).

Numerical values of critical lengths of a strip for the above stated class of polymer material under the action of oily corrosive medium, are given in Table 1.9. The data of thrust stresses $\sigma(\lambda)$ were taken from Table 1.3.

Thus, the suggested stability criterion (1.144) and (1.145) for structural elements maintained in the corrosive medium, and also the suggested fundamental results on strength of structural elements allow creating a principally new engineering calculation methods for a whole class of unknown before problems of industrial constructions operating in liquid media.

KEYWORDS

- **Corrosive liquid**
- **Fourier criterion**
- **Interdiffusion**
- **Onzager linear equations**
- **Thermodynamical force**

CHAPTER 2

STRESS–STRAIN STATE OF A SANDWICH THICK-WALLED PIPE WITH REGARD TO CHANGE OF PHYSICO-CHEMICAL PROPERTIES OF THE MATERIAL

G. G. ALIYEV and F. B. NAGIYEV

CONTENTS

2.1 STRESS, STRAIN, AND STRENGTH OF A SANDWICH PIPE MADE UP OF A POLYMER MATERIAL SITUATED UNDER THE ACTION OF CORROSIVE LIQUID MEDIUM AND EXTERNAL LOADS

We investigate the strength problem of a sufficiently long sandwich pipe with regard to physico–chemical properties of the layers, that arise when contacting with corrosive medium. Each layer is assumed to be piecewise homogeneous with mechanical properties $\sum_n(\lambda_n)$, $\nu_n(\lambda_n)$, $G_n(\lambda_n)$ and physico–chemical changes λ_n, α_n, $\phi_n(\lambda_n)$, $\psi_n(\lambda_n)$, $\eta_n(\lambda_n)$. Let the pipe consist of $(N-1)$ piecewise homogeneous layers and (N) boundary surfaces. Let the sandwich pipe be under the action of axial tensile strength P, of internal p_a and external p_b pressures. The goal of paper is to determine the stress and strain quantities in layers and also contact stresses between the layers that determine adhesive strength depending both on external forces and on degree of change of physico–chemical properties of layers, that is on λ_n, α_n, $\varphi_n(\lambda_n)$, $\psi_n(\lambda_n)$, $\eta_n(\lambda_n)$, and also on the effect of thrust mass forces arising in layers of the sandwich pipe. Problem is solved under total cohesion between the layers (Aliyev, 1995, 1998, 2012, , 2012).

Consider the nth layer of the binder ($r_n \leq r \leq r_{n+1}$) in the curvilinear cylindrical system of coordinates and stress σ_{zz}, σ_{yy}, σ_{rr}, and strain ε_{zz}, ε_{yy}, ε_{rr} (Figure 2.1) components corresponding to these layers.

Since this hell's structure is symmetric with respect to the axis and its length is significantly greater than the sizes of cross-section, then the calculation may be based on the following conjecture of generalized plane deformation: the lateral cross sections of the pipe that were plane before deformation remain plane after deformation for all values of the swelling parameter λ_n, too, that is, the axial deformation ε_z is constant in cross-section $z = const$; all stresses and deformations for all values of the swelling parameter λ_n are independent of the coordinate z if P, p_a, p_b are constant along the shell's length or are slowly changing with respect to z functions. Then for any elastic layer we can use the relations:

$$\varepsilon_{rr} = \frac{dw}{dr}, \ \varepsilon_{yy} = \frac{w}{r}, \ \varepsilon_{zz} = \varepsilon = const$$

$$\frac{\partial \varepsilon_{zz}}{\partial r} = 0, \; w = w(r) \tag{2.1}$$

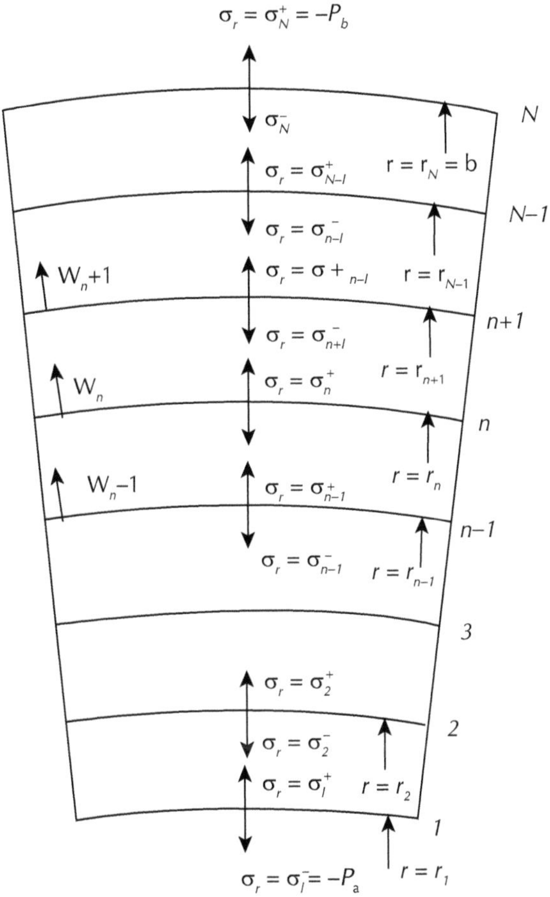

FIGURE 2.1 Multi-layer pipe made up of polymer material

The Hooke's generalized law (1.91) with regard to correction factors $\varphi_n(\lambda_n)$, $\psi_n(\lambda_n)$, $\eta_n(\lambda_n)$ (1.92) for the points of each n th layer ($r_n \le r \le r_{n+1}$) will have the form (Aliyev, 2012):

$$\begin{cases} \sigma_{ij} = a_{on}\psi_n(\lambda_n)\cdot\theta\cdot\delta_{ij} + 2G_{on}\phi_n(\lambda_n)\varepsilon_{ij} - \eta_{on}\cdot\eta_n(\lambda_n)\cdot\alpha_n\cdot\tilde{\lambda}_n\cdot\delta_{ij} \\ \sigma(\varepsilon,\lambda_n) = 3a_{0n}\psi_n(\lambda_n)[1+\dfrac{2G_{0n}}{3a_{0n}}\cdot\dfrac{\phi_n(\lambda_n)}{\psi_n(\lambda_n)}](\theta - 3\tilde{\alpha}_n\tilde{\lambda}_n) \end{cases} \tag{2.2}$$

In the expanded form:

$$\begin{cases} \sigma_{rr} = a_{on}\psi_n(\lambda_n)\cdot\theta + 2G_{on}\phi_n(\lambda_n)\varepsilon_{rr} - \eta_{on}\eta_n(\lambda_n)\alpha_n\tilde{\lambda}_n \\ \sigma_{yy} = a_{on}\psi_n(\lambda_n)\cdot\theta + 2G_{on}\phi_n(\lambda_n)\varepsilon_{yy} - \eta_{on}\eta_n(\lambda_n)\alpha_n\tilde{\lambda}_n \\ \sigma_{zz} = a_{on}\psi_n(\lambda_n)\cdot\theta + 2G_{on}\phi_n(\lambda_n)\varepsilon_{zz} - \eta_{on}\eta_n(\lambda_n)\alpha_n\tilde{\lambda}_n \end{cases} \tag{2.3}$$

$$\sigma(\varepsilon,\lambda_n) = 3a_{0n}\psi_n(\lambda_n)[1+\frac{2G_{0n}}{3a_{0n}}\cdot\frac{\phi_n(\lambda_n)}{\psi_n(\lambda_n)}](\theta - 3\tilde{\alpha}_n\tilde{\lambda}_n) \tag{2.4}$$

Here σ_{ij} and ε_{ij} are stress and deformations tensors: $\theta = \varepsilon_{ii} = \varepsilon_{rr} + \varepsilon_{yy} + \varepsilon_{zz}$ is a cubic deformation;

$$a_{on} = \frac{E_{on}\nu_{on}}{(1+\nu_{on})(1-2\nu_{on})},\quad 2G_{on} = \frac{E_{on}}{1+\nu_{on}} \tag{2.5}$$

are the Lame coefficients for the nth layer regardless of swelling, that is, $\lambda_n = 0$; λ_n is a swelling parameter of nth layer of polymer material; α_n is a linear swelling factor of nth layer; the correction factors $\varphi_n(\lambda_n)$, $\psi_n(\lambda_n)$, and $\eta_n(\lambda_n)$, dependent on swelling parameter of the nth layer λ_n of pipe have form of (1.86):

$$\phi_n(\lambda) = \frac{b_n^{\tilde{\lambda}_n}}{1+\dfrac{k_n}{1+\nu_{on}}\tilde{\lambda}_n},\quad \psi_n(\lambda) = \phi_n(\lambda)\frac{1+\dfrac{k_n}{\nu_{on}}\tilde{\lambda}_n}{1-\dfrac{2k_n}{1-2\nu_{0n}}\tilde{\lambda}_n},$$

$$\eta_n(\lambda) = \frac{b_n^{\tilde{\lambda}} \lambda_{n\max}}{1 - \frac{2k_n}{1-2\nu_{on}} \tilde{\lambda}_n}, \; \eta_{0n} = \frac{E_{0n}}{1-2\nu_{0n}} \tag{2.6}$$

$$\tilde{\lambda}_n = \frac{\lambda_n}{\lambda_{n\max}}, \; k_n = \nu_{n\max} - \nu_{on}, \; b_n = \frac{E_{n\min}}{E_{on}}, \; \tilde{\alpha}_n = \alpha_n \lambda_{\max} \tag{2.7}$$

E_{0n} and ν_{0n} is elasticity modulus and Poisson's ratio respectively, for the nth layer regardless of swelling, that is, for $\lambda_n = 0$. Dependence of volumetric stress $\sigma = \sigma_{ii} = \sigma_{rr} + \sigma_{yy} + \sigma_{zz}$ and total cubic deformation in the nth layer ($r_n \le r \le r_{n+1}$), according to (1.92) will equal:

$$\sigma = 3a_{on}\psi_n(\lambda_n)[1 + \frac{2G_{on}}{3a_{on}} \frac{\phi_n(\lambda_n)}{\psi_n(\lambda_n)}](\theta - 3\tilde{\alpha}_n \lambda_n) \tag{2.8}$$

Differential equilibrium equation of elementary volume of the nth layer in radial direction will be ($r_n \le r \le r_{n+1}$):

$$\frac{d\sigma_{rr}}{dr} = \frac{\sigma_{yy} - \sigma_{rr}}{r} \tag{2.9}$$

Substituting (2.3) and (2.4) in (2.9) and taking into account (2.1) and the expression $\varepsilon_{yy} = \frac{\partial(r\varepsilon_{yy})}{\partial r}$, the equilibrium equation of elementary nth layer will be reduced to the differential equation with respect to deflection $w(\lambda)$ of the form:

$$\frac{dw}{dr} + \frac{w}{r} = c_n(\lambda) \tag{2.10}$$

The common integral of equation (2.10) will be:

$$w(\lambda) = Ar + \frac{B}{r} \tag{2.11}$$

Then allowing for (2.1), deformations (2.11) will take the form:

$$\varepsilon_{yy} = \frac{w}{r} = A + \frac{B}{r^2},\ \varepsilon_{rr} = \frac{\partial w}{\partial r} = A - \frac{B}{r^2},\ \varepsilon_{zz} = \varepsilon = const \tag{2.12}$$

The constants A and B are determined by boundary conditions for the nth layer of the sandwich pipe ($r_n \le r \le r_{n+1}$):

$$w = w_n \ , \ w = w_n$$

$$r = r_{n+1},\ w = w_{n+1} \tag{2.13}$$

where, w_n and w_{n+1} are deflections of contour points of the nth layer. Introduce the denotation:

$$\varepsilon_{yy} = \frac{w}{r}\Big|_{r=r_n} = \varepsilon_n,\ \varepsilon_{yy} = \frac{w}{r}\Big|_{r=r_{n+1}} = \varepsilon_{n+1} \tag{2.14}$$

Satisfying condition (2.13) allowing (2.11) and (2.12), the constants A and B will be expressed by the contour values of deflections w_n and w_{n+1} in the form:

$$A = \frac{(wr)_{n+1} - (wr)_n}{r_{n+1}^2 - r_n^2},\ B = \frac{(w/r)_{n+1} - (w/r)_n}{r_{n+1}^{-2} - r_n^{-2}} \tag{2.15}$$

Then the deformations of the points of the nth layer (2.12) ($r_n \le r \le r_{n+1}$) will be expressed by the contour values of deformations ε_n and ε_{n+1} in the form:

$$\begin{pmatrix} \varepsilon_{yy} \\ \varepsilon_{rr} \end{pmatrix} = \frac{\varepsilon_{n+1} r_{n+1}^2 - \varepsilon_n r_n^2}{r_{n+1}^2 - r_n^2} \pm \frac{1}{r^2} \cdot \frac{\varepsilon_{n+1} - \varepsilon_n}{r_{n+1}^{-2} - r_n^{-2}} \tag{2.16}$$

Substituting (2.16) and (2.5) in (2.3), define the stress at the points of the nth layer of a sandwich layer on the thickness $r_n \leq r \leq r_{n+1}$ with regard to change of physico–chemical properties of material of the nth layer of pipe λ_n in the form:

$$\frac{1}{2G_{on}\phi_n(\lambda_n)}\begin{pmatrix}\sigma_{yy}\\ \sigma_{rr}\end{pmatrix} = \frac{\nu_{on}}{1-2\nu_{on}}\frac{\psi_n(\lambda_n)}{\phi_n(\lambda_n)}\varepsilon_z +$$

$$[1+\frac{2\nu_{on}}{1-2\nu_{on}}\cdot\frac{\psi_n(\lambda_n)}{\phi_n(\lambda_n)}]\frac{\varepsilon_{n+1}r_{n+1}^2-\varepsilon_n r_n^2}{r_{n+1}^2-r_n^2} \pm$$

$$\frac{1}{r^2}\cdot\frac{\varepsilon_{n+1}-\varepsilon_n}{r_{n+1}^{-2}-r_n^{-2}} - \frac{\eta_{on}}{2G_{on}}\frac{\eta_n(\lambda_n)}{\phi_n(\lambda_n)}\alpha_n\tilde{\lambda}_n \quad (2.17)$$

(for $n = 1,2,\ldots,N-1$)

$$\frac{1}{2G_{0n}\phi_n(\lambda_n)}\sigma_{zz} = [1+\frac{\nu_{on}}{1-2\nu_{on}}\cdot\frac{\psi_n(\lambda_n)}{\phi_n(\lambda_n)}]\varepsilon_z +$$

$$\frac{2\nu_{on}}{1-2\nu_{on}}\frac{\psi_n(\lambda_n)}{\phi_n(\lambda_n)}\frac{\varepsilon_{n+1}r_{n+1}^2-\varepsilon_n r_n^2}{r_{n+1}^2-r_n^2} - \quad (2.18)$$

$$\frac{\eta_{on}}{2G_{on}}\frac{\eta_n(\lambda_n)}{\phi_n(\lambda_n)}\cdot\alpha_n\cdot\tilde{\lambda}_n$$

Then from (2.17), radial stresses on boundary surfaces $\sigma_r|_{r=r_n} = \sigma_n^+$ and $\sigma_r|_{r=r_{n+1}} = \sigma_{n+1}^-$ of the nth layer of a sandwich pipe will equal:

$$\frac{1}{2G_{on}\phi_n(\lambda_n)}\sigma_n^+ = \frac{\nu_{on}}{1-2\nu_{on}}\cdot\frac{\psi_n(\lambda_n)}{\phi_n(\lambda_n)}\varepsilon_z - \frac{1}{r_{n+1}^2-r_n^2}\{r_{n+1}^2 +$$

$$+[1+\frac{2\nu_{on}}{1-2\nu_{on}}\cdot\frac{\psi_n(\lambda_n)}{\phi_n(\lambda_n)}]r_n^2\}\cdot\varepsilon_n +$$

$$+\frac{2r_{n+1}^2}{r_{n+1}^2-r_n^2}[1+\frac{\nu_{on}}{1-2\nu_{on}}\frac{\psi_n(\lambda_n)}{\phi_n(\lambda_n)}]\varepsilon_{n+1}-$$
$$-\frac{\eta_{on}}{2G_{on}}\frac{\eta_n(\lambda_n)}{\phi_n(\lambda_n)}\cdot\alpha_n\cdot\tilde{\lambda}_n \qquad (2.19)$$

(for n=1,2,…,N-1)

$$\frac{1}{2G_{on}\phi_n(\lambda_n)}\sigma_{n+1}^-=\frac{\nu_{on}}{1-2\nu_{on}}\cdot\frac{\psi_n(\lambda_n)}{\phi_n(\lambda_n)}\varepsilon_z-$$
$$\frac{2r_n^2}{r_{n+1}^2-r_n^2}[1+\frac{\nu_{on}}{1-2\nu_{on}}\cdot\frac{\psi_n(\lambda_n)}{\phi_n(\lambda_n)}]\varepsilon_n+$$
$$+\frac{1}{r_{n+1}^2-r_n^2}[r_n^2+(1+\frac{2\nu_{on}}{1-2\nu_{on}}\cdot\frac{\psi_n(\lambda_n)}{\phi_n(\lambda_n)})r_{n+1}^2]\varepsilon_{n+1}-$$
$$\frac{\eta_{on}}{2G_{on}}\frac{\eta_n(\lambda_n)}{\phi_n(\lambda_n)}\cdot\alpha_n\cdot\tilde{\lambda}_n \qquad (2.20)$$

(for n=1,2,…,N-1)

Define the adhesive strength of a sandwich pipe on the *n*th boundary surface with regard to change of physico–chemical properties of layers in the form of equality to zero of difference of radial stresses σ_n^+ and σ_n^- acting on the *n*th layer of boundary surface, in the form (Aliyev, 1987): $\Delta\sigma_n=\sigma_n^+-\sigma_n^-=0$, (for n=1,2,…,N) (2.21).

From (2.20) define σ_n^-. For that in (2.20) replace index (*n*+1) by *n*, and get:

$$\frac{1}{2G_{on-1}\phi_{n-1}(\lambda_{n-1})}\sigma_n^-=\frac{\nu_{on-1}}{1-2\nu_{on-1}}\cdot\frac{\psi_{n-1}(\lambda_{n-1})}{\phi_{n-1}(\lambda_{n-1})}\varepsilon_z-\frac{2r_{n-1}^2}{r_n^2-r_{n-1}^2}[1+$$
$$\frac{\nu_{on-1}}{1-2\nu_{on-1}}\cdot\frac{\psi_{n-1}(\lambda_{n-1})}{\phi_{n-1}(\lambda_{n-1})}]\varepsilon_{n-1}+\frac{1}{r_n^2-r_{n-1}^2}[r_{n-1}^2+(1+\frac{2\nu_{o\,n-1}}{1-2\nu_{o\,n-1}}\cdot\frac{\psi_{n-1}(\lambda_{n-1})}{\varphi_{n-1}(\lambda_{n-1})})r_n^2]\varepsilon_n-$$
$$\frac{\eta_{on-1}}{2G_{on-1}}\frac{\eta_{n-1}(\lambda_{n-1})}{\phi_{n-1}(\lambda_{n-1})}\alpha_{n-1}\tilde{\lambda}_{n-1} \qquad (2.22)$$

(for n = 2,…, N-1)

Using (2.19) and (2.22), the contact condition on the *n*th boundary surface (2.21) will have the form:

$$4G_{on}\phi_n(\lambda_n)\frac{r_{n+1}^2}{r_{n+1}^2-r_n^2}(1+\frac{\nu_{on}}{1-2\nu_{on}}\frac{\psi_n(\lambda_n)}{\phi_n(\lambda_n)})\varepsilon_{n+1}$$

$$-\{\frac{2G_{on}\phi_n(\lambda_n)}{r_{n+1}^2-r_n^2}\cdot[r_{n+1}^2+(1+\frac{2\nu_{on}}{1-2\nu_{on}}\cdot\frac{\psi_n(\lambda_n)}{\phi_n(\lambda_n)})r_n^2]+$$

$$\frac{2G_{on-1}\phi_{n-1}(\lambda_{n-1})}{r_n^2-r_{n-1}^2}[r_{n-1}^2+(1+\frac{2\nu_{on-1}}{1-2\nu_{on-1}}\cdot\frac{\psi_{n-1}(\lambda_{n-1})}{\phi_{n-1}(\lambda_{n-1})})r_n^2]\}\varepsilon_n+$$

$$4G_{on-1}\phi_{n-1}(\lambda_{n-1})\frac{r_{n-1}^2}{r_n^2-r_{n-1}^2}[1+\frac{\nu_{on-1}}{1-2\nu_{on-1}}\cdot\frac{\psi_{n-1}(\lambda_{n-1})}{\phi_{n-1}(\lambda_{n-1})}]\varepsilon_{n-1}+$$

$$2[\frac{G_{on}\nu_{on}}{1-2\nu_{on}}\cdot\psi_n(\lambda_n)-\frac{G_{on-1}\nu_{on-1}}{1-2\nu_{on-1}}\cdot\psi_{n-1}(\lambda_{n-1})]\varepsilon_z$$

$$=\eta_{on}\eta_n(\lambda_n)\alpha_n\tilde{\lambda}_n-\eta_{on-1}\eta_{n-1}(\lambda_{n-1})\alpha_{n-1}\tilde{\lambda}_{n-1} \qquad (2.23)$$

(for $n=2,\ldots,N-1$).

The values of radial stresses of the first σ_1^- and last σ_N^+ layers are determined from boundary conditions:

$$r=r_1,\ \sigma_1^-=-p_a \qquad (2.24)$$

$$r = r_N\,,\ \sigma_N^+ = -p_b \tag{2.25}$$

From (2.21), the contact condition on the first and the *N*th boundary surface will be in the following form:

$$\Delta\sigma_1 = \sigma_1^+ - \sigma_1^- = 0 \tag{2.26}$$

$$\Delta\sigma_N = \sigma_N^+ - \sigma_N^- = 0 \tag{2.27}$$

From (2.26) and (2.27), allowing (2.24) and (2.25), define σ_1^+ and σ_N^-, in the form:

$$\sigma_1^+ = -p_a \tag{2.28}$$

$$\sigma_N^- = -p_b \tag{2.29}$$

From (2.19) for $n = 1$ and (2.28) we get a contact condition on the first boundary surface, in the form:

$$\frac{2r_2^2}{r_2^2 - r_1^2}\left(1 + \frac{\nu_{01}}{1-2\nu_{01}}\frac{\psi_1(\lambda_1)}{\phi_1(\lambda_1)}\right)\varepsilon_2 - \frac{1}{r_2^2 - r_1^2}[r_2^2 +$$
$$\left(1 + \frac{2\nu_{01}}{1-2\nu_{01}}\cdot\frac{\psi_1(\lambda_1)}{\phi_1(\lambda_1)}\right)r_1^2]\varepsilon_1 +$$

$$\frac{\nu_{01}}{1-2\nu_{01}}\frac{\psi_1(\lambda_1)}{\phi_1(\lambda_1)}\varepsilon_z = \frac{1}{2G_{01}\phi_1(\lambda_1)}[\eta_{01}\eta_1(\lambda_1)\alpha_1\tilde{\lambda}_1 - p_a] \tag{2.30}$$

From (2.20) for $n = N - 1$ and (2.29) we get a contact condition on the *N*th boundary surface, in the form:

$$\frac{1}{r_N^2 - r_{N-1}^2}[r_{N-1}^2 + (1 + \frac{2\nu_{0N-1}}{1-2\nu_{ON-1}} \cdot \frac{\psi_{N-1}(\lambda_{N-1})}{\phi_{N-1}(\lambda_{N-1})})r_N^2]\varepsilon_N -$$

$$\frac{2r_{N-1}^2}{r_N^2 - r_{N-1}^2} \cdot [1 +$$

$$\frac{\nu_{0N-1}}{1-2\nu_{0N-1}} \cdot \frac{\psi_{N-1}(\lambda_{N-1})}{\phi_{N-1}(\lambda_{N-1})}]\varepsilon_{N-1} + \frac{\nu_{ON-1}}{1-2\nu_{ON-1}} \cdot \frac{\psi_{N-1}(\lambda_{N-1})}{\phi_{N-1}(\lambda_{N-1})}\varepsilon_{zz} =$$

$$\frac{1}{2G_{0N-1}\phi_{N-1}(\lambda_{N-1})}[\eta_{0N-1} \cdot \eta_{N-1}(\lambda_{N-1}) \cdot \alpha_{N-1} \cdot \tilde{\lambda}_{N-1} - p_b] \quad (2.31)$$

The longitudinal force P in a thick-walled sandwich pipe equals:

$$P = \pi \sum_{n=1}^{N-1} (r_{n+1}^2 - r_n^2)\sigma_{zz} \quad (2.32)$$

Substituting (2.18) in (2.32) we get:

$$4\sum_{n=1}^{N-1} G_{0n}\psi_n(\lambda_n)\frac{\nu_{0n}}{1-2\nu_{0n}}(\varepsilon_{n+1}r_{n+1}^2 - \varepsilon_n r_n^2) +$$

$$2\varepsilon_z \cdot \sum_{n=1}^{N-1} G_{on}\phi_n(\lambda_n)(r_{n+1}^2 - r_n^2)(1 + \frac{\nu_{on}}{1-2\nu_{on}} \frac{\psi_n(\lambda_n)}{\phi_n(\lambda_n)})$$

$$= \frac{1}{\pi}P + \sum_{n=1}^{N-1} \eta_{on}\eta_n(\lambda_n)(r_{n+1}^2 - r_n^2)\alpha_n\tilde{\lambda}_n \quad (2.33)$$

We note the following: In the system of algebraic equations (2.30), (2.23), (2.31), and (2.33), there are swelling parameters λ_n and λ_{n-1} corresponding to the heterogeneous polymer materials. It is obvious that character of change of physico–chemical property of each of polymer materials will be different for the same period of their stay in corrosive liquid medium. In this connection, the necessary mathematical condition of relation between these swelling parameters is suggested by formula (1.57).

Therefore, by finding the solution of algebraic system (2.30), (2.23), (2.31), and (2.33) in the place of mathematical relation between swelling parameters of the *n*th and (*n*–1)th layers of a sandwich pipe, we should use a condition of form (Aliyev, 2012, 2012):

$$\lambda_n = \frac{q_n}{q_{n-1}} \lambda_{n-1} \tag{2.34}$$

Here q_n and q_{n-1} are linear coefficients in dependencies of swelling parameter λ_n on time *t* in the *n*th and (*n*–1)th layers of a sandwich pipe.

Thus, the system of equations (2.30), (2.23), (2.31), and (2.33) is a (*N*+1)linear-algebraic equation with same number of unknown tangential $\varepsilon_1, \varepsilon_2, \ldots, \varepsilon_N$ and longitudinal ε_{zz} deformations with regard to physico–chemical changeability of pipe's layers and thrust mass forces effect arising in layers. If the solution of system of equations was found with respect to external loads P, p_a, p_b and swelling parameters λ_n, then by formula (2.17) and (2.18), it is easy to determine the stress distribution on thickness of pipe's layers and also adhesive strength, that is, maximal breaking off stress between the layers by formula:

$$(\sigma_{br.off})_{\max} = \max\begin{pmatrix} \sigma_n^+ \\ \sigma_n^- \end{pmatrix} \tag{2.35}$$

Special case: While considering a special case of sandwich pipe under the action of only swelling effect, in algebraic system (2.30), (2.23), (2.31), and (2.33), it will be necessary to equate the external mechanical loads P, p_a, p_b to zero. This case is interesting in studying the character of stress and deformation distribution because of physico–chemical change of layer's character. Therewith, it is interesting to know maximal breaking off stresses on boundaries between the layers (2.35) depending on the change of physico–chemical properties of layers material and thrust mass forces arising in a sandwich pipe.

2.2 ON A CRITERION OF ADHESIVE STRENGTH OF SANDWICH AND REINFORCED BODIES WITH REGARD TO PHYSICO–CHEMICAL CHANGE OF LAYER'S MATERIAL

By designing sandwich pipes from polymer and composite materials, one of the main requirements to reliability of construction's operation is provision of continuity on interphase of layers, that is, impermissibility of interlayer lamination. The adhesive force phenomena belong to the complicated phenomena of nature. Nature of adhesive force between two materials is the result of synthesis of a number of physico–chemical and physico–mechanical phenomena. Firstly, as a result of application of finish (i.e., adhesive interlayer) on the boundary of contact of two material, there arise intermolecular adhesion forces called Van der Waals forces; secondly, by shaping the goods by compaction pressure method and temperature, there arises mechanico–chemical kind of forces; thirdly, the existing natural roughness of surface on the interface of two polymer materials creates the adhesive forces of purely mechanical character, and so on. Today, there is no fundamental formula that could establish acceptable dependence of adhesive forces on characteristics of physico–chemical and physico–mechanical changeability of the material. In engineering practice, the admissible quantity of adhesive force is established by special experiments.

In classic approach, in place of common criterion of adhesive strength, stress intensity σ_i (or deformation intensity ε_i) is used at the points of surface contact of two bodies, that does not increase the pregiven quantity of breaking off stress called critical value of adhesive stress or adhesive strength:

$$\sigma_i = (\sigma_{ij}\sigma_{ij})^{1/2} \leq \sigma_{br.off} \text{ or } \varepsilon_i \leq (\varepsilon_i)_{br.off} \tag{2.36}$$

In simplified problems, this criterion may have a simpler form. For example, in sandwich cylindrical pipes, there may be used a condition imposed on radial or tangential interlayer stress (or deformation). In this case, the radial or tangential stress (deformation) on the contact boundary must not exceed the pregiven number of breaking off values of the material of form:

$$\sigma_{rr} \le \sigma_{rr}^{br.off} \text{ or } \varepsilon_{rr} \le \varepsilon_{rr}^{br.off} \quad (2.37)$$

These classic criteria of material strength are not suitable for the materials and constructions made of them, because these or other causes change in their physico–chemical or physico–mechanical properties. For example, to materials whose physico–chemical or physico–mechanical properties are sensible to aggressive media effects or different type actions of electromagnetic character and so on.

In this connection (in this section), we suggest a criterion of interlayer breaking off with regard to change of physico–mechanical properties of intercontacting materials and of finish (i.e., adhesive interlayer) in form:

$$\sigma_i \le \sigma_i^{br.off}(\lambda_k) \quad (2.38)$$

Here: λ_K are parameters defining physico–chemical changeability of contacting layers of construction; $\sigma_i^{br.off}(\lambda_k)$ is the breaking off stress of layers dependent on the character of physico–chemical changeability. It is suggested to define the form of function $\sigma_i^{br.off}(\lambda_k)$ in the following way. Under the action of physico–chemical changeability of polymer material, the mechanical properties of a sandwich body will change proportional to the swelling parameter λ_n of the layers material. As a rule, this effect reduces to weaken the force coupling between the layers, that is, reduces to change the adhesive strength between the layers (Figure 2.2). In this connection, the form of ultimate stress in interlayer breaking off, that is, of adhesive force, is suggested in the form [9,12]:

$$\sigma_{br.off}(\lambda) = Ae^{-\alpha\lambda} \quad (2.39)$$

Here λ is the swelling parameter of polymer material (Figure 2.2).

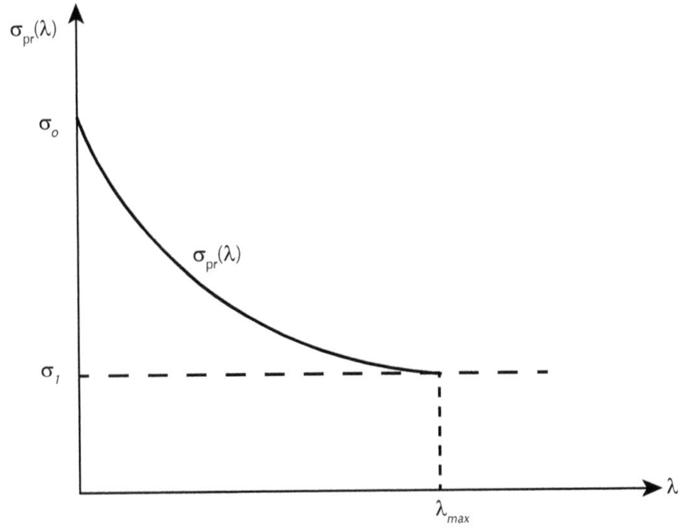

FIGURE 2.2 Dependence of ultimate stress at interlayer breaking off on physico–chemical changeability of polymer material

For defining the constants A and α, we use the following conditions:
For $\lambda = 0$, $\sigma_{br.off}(0) = \sigma_0$

For $$\lambda = \lambda_{\max},\ \sigma_{br.off}(\lambda_{\max}) = \sigma_1 \tag{2.40}$$

The coefficients A and α are defined from (2.39) and (2.40) in the form:

$$A = \sigma_0,\ \alpha = -\frac{1}{\lambda_{\max}} e^{\ln\frac{\sigma_1}{\sigma_0}} \tag{2.41}$$

Substituting (2.41) in (2.39), we get dependence of breaking off stress $\sigma_{br.off}$ for each value of the swelling parameter λ of polymer material, changing on the interval $0 \le \lambda \le \lambda_{\max}$, on limit values of stresses σ_0 and σ_1, in the form:

$$\sigma_{br.off} = \sigma_0 \left(\frac{\sigma_1}{\sigma_0}\right)^{\frac{\lambda}{\lambda_{\max}}} \tag{2.42}$$

Here $\sigma_{br.off}(0)=\sigma_0$ is the ultimate strength quantity of the material for the case regardless of physico–chemical change in polymer material, that is, for $\sigma_{br.off}(0)=\sigma_0$ is the breaking off stress between the layers with regard to physico–chemical change of the material, that is, for $\lambda=0$; $\sigma_{br.off}(\lambda_{\max})=\sigma_1$. These quantities are experimentally determined. Thus, if the ultimate strength σ_0 and the quantity σ_1 of the breaking off stress λ with regard to physico–chemical change of the material are determined experimentally, then the breaking off stress $\sigma_{br.off}(\lambda)$ for each value of the interval $0\le\lambda\le\lambda_{\max}$ will be determined by formula (2.42). In this case, in place of adhesive strength criterion of polymer material with regard to effect of physico–chemical change of material, we suggest the following (Aliyev, 1980):

$$\sigma_{br.off}(\lambda)\le\sigma_0\left(\frac{\sigma_1}{\sigma_0}\right)^{\frac{\lambda}{\lambda_{\max}}},\ \text{for}\ (0\le\lambda\le\lambda_{\max}) \qquad (2.43)$$

Example: Give numerical calculation of dependence of limit values of stress on swelling parameter λ, changing on the interval $(0\le\lambda\le\lambda_{\max})$. For that use formula (2.43) of the form:

$$\sigma_{br.off}(\lambda)\le\sigma_0\left(\frac{\sigma_1}{\sigma_0}\right)^{\frac{\lambda}{\lambda_{\max}}}\quad \text{or}\ (0\le\lambda\le\lambda_{\max})$$

From Figure 1.6 of chapter 1, σ_0, σ_1 and $\lambda_{\max}$ are defined for a class of polymer material, represented in Table 2.1.

TABLE 2.1 σ_0, σ_1 and $\lambda_{\max}$ for a class of polymer material

Brand	$\lambda_{\max}$	σ_0 (MPa)	σ_1 (MPa)	*Model* $\sigma_{br.off}(\lambda)\le\sigma_0\left(\frac{\sigma_1}{\sigma_0}\right)^{\lambda/\lambda_{\max}}$
PS	0.02	20	15.6	$\sigma_{br.off}(\lambda)\le 20\times 0.78^{50\cdot\lambda}$
PA	0.028	56.6	45.9	$\sigma_{br.off}(\lambda)\le 56.6\times 0.81^{35.714\cdot\lambda}$

TABLE 2.1 *(Continued)*

PP	0.02	36.9	31.6	$\sigma_{br.off}(\lambda) \le 36.9 \times 0.86^{50 \cdot \lambda}$
PEHD	0.0275	27.9	22.9	$\sigma_{br.off}(\lambda) \le 27.9 \times 0.82^{36.364 \cdot \lambda}$
PELD	0.035	15.6	10.7	$\sigma_{br.off}(\lambda) \le 15.6 \times 0.69^{28.571 \cdot \lambda}$

Limit values of the breaking off stress will be in form:

$$\sigma_0 \ge \sigma_{br.off}(\lambda) = \sigma_0 \left(\frac{\sigma_1}{\sigma_0}\right)^{\lambda / \lambda_{\max}} \ge \sigma_1$$

The following areas of change of breaking off values of stresses $\sigma_{br.off}(\lambda)$ (of adhesive stresses) for swelling parameter λ , changing on the interval $0 \le \lambda \le \lambda_{\max}$ are established below by numerical calculations:

(1) for PS: $20 \ge \sigma_{br.off}(\lambda) = 20 \times 0.78^{50 \cdot \lambda} \ge 15.6$

(2) for PA: $56.6 \ge \sigma_{br.off}(\lambda) = 56.6 \times 0.81^{35.714 \cdot \lambda} \ge 45.9$

(3) for PP: $36.9 \ge \sigma_{br.off}(\lambda) = 36.9 \times 0.86^{50 \cdot \lambda} \ge 31.6$

(4) for PEHD: $27.9 \ge \sigma_{br}(\lambda) = 27.9 \times 0.82^{36,364 \cdot \lambda} \ge 22.9$

(5) for PELD: $15.6 \ge \sigma_{br.off}(\lambda) = 15.6 \times 0.69^{28.571 \cdot \lambda} \ge 10.7$

Therewith, note that formula (2.43) is suitable for the case of metal-polymer. In other words, when one of the materials is not sensitive to physico–chemical changes. But in the case of a sandwich body in the form of polymer-polymer, the adhesive strength with regard to physico–chemical change of both materials will be defined in the following way. On the basis of (2.43), for each material of the contiguous layer of sandwich layer, we can write the adhesive strength with regard to physico–chemical change of both materials, in the form:

$$\begin{cases} \sigma_{1br.off}(\lambda_1) \le \sigma_{01}\left(\frac{\sigma_1}{\sigma_{01}}\right)^{\frac{\lambda_1}{\lambda_{1\max}}} \\ \sigma_{2br.off}(\lambda_2) \le \sigma_{02}\left(\frac{\sigma_2}{\sigma_{02}}\right)^{\frac{\lambda_2}{\lambda_{2\max}}} \end{cases} \quad \begin{matrix} (0 \le \lambda_1 \le \lambda_{1\max}) & (2.44) \\ \text{for } (0 \le \lambda_2 \le \lambda_{2\max}) & (2.45) \end{matrix}$$

It was established above by formula (2.34) ,that for one and the same time, the coupling between the swelling parameters is of form:

$$\lambda_2 = \frac{q_2}{q_1}\lambda_1, \quad \lambda_{2\max} = \frac{q_2}{q_1}\lambda_{1\max} \tag{2.46}$$

Substituting (2.46) to (2.45), we find breaking off stress for the second polymer material depending on the swelling parameter λ_1 of the first polymer layer:

$$\sigma_{2br.off}(\lambda_2) \le \sigma_{02}\left(\frac{\sigma_2}{\sigma_{02}}\right)^{\frac{\lambda_1}{\lambda_{1\max}}} \tag{2.47}$$

Thus, the adhesive strength criterion, that is, the breaking off stress of the first and second layers will be represented depending on the swelling parameters of the first layer λ_1 in form:

$$\begin{cases} \sigma_{1br.off}(\lambda_1) \le \sigma_{01}\left(\frac{\sigma_1}{\sigma_{01}}\right)^{\frac{\lambda_1}{\lambda_{1\max}}} \\ \sigma_{2br.off}(\lambda_2) \le \sigma_{02}\left(\frac{\sigma_2}{\sigma_{02}}\right)^{\frac{\lambda_1}{\lambda_{1\max}}} \end{cases} \quad \text{for } (0 \le \lambda_1 \le \lambda_{\max}) \tag{2.48}$$

2.3 STRESS–STRAIN STATE OF TWO-LAYER PIPE SITUATED UNDER THE ACTION OF CORROSIVE LIQUID MEDIUM AND EXTERNAL LOADS

Investigate the strength problem of two-layer pipe consisting of polymer materials with regard to change of physico–chemical properties of layers and also of arising thrust mass forces in the layers of a sandwich pipe, that arises after contacting with corrosive medium (Figure 2.3).

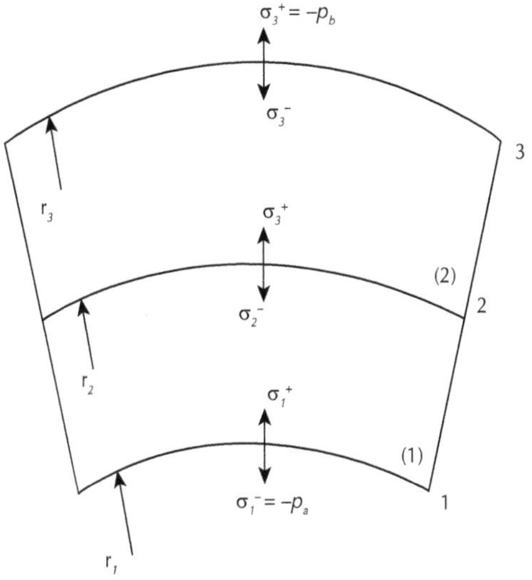

FIGURE 2.3 Two-layer pipe made up of polymer material

It is supposed that the pipe's layers are pricewise homogeneous with the following mechanical and physico–mechanical properties:

Layer I— E_{01}, ν_{01}, G_{01}, η_{01}, λ_1, α_1, $\varphi_1(\lambda_1)$, $\psi_1(\lambda_1)$, and $\eta_1(\lambda_1)$

Layer II— E_{02}, ν_{02}, G_{02}, η_{02}, λ_2, α_2, $\varphi_2(\lambda_2)$, $\psi_2(\lambda_2)$, and $\eta_2(\lambda_2)$ (2.49)

Define stresses and deformations in the layers and maximal breaking off stresses between the layers, that define adhesive strength of a sandwich pipe situated under the joint action of external loads p_a, p_b, P and swelling forces λ_n of the layers material. The problem's solution will be determined by means of general equations obtained in previous sections.

For that, in the system of equations (2.30), (2.23), (2.31), and (2.33) accept $n = 2$. The problem is reduced to the system of four algebraic equations with respect to unknown contour values of tangential ε_1, ε_2, ε_3 and longitudinal $\varepsilon_z = \varepsilon_4$ deformations of the form:

$$-\frac{1}{r_2^2 - r_1^2}[r_2^2 + (1 + \frac{2\nu_{01}}{1-2\nu_{01}}\frac{\psi_1(\lambda_1)}{\phi_1(\lambda_1)})r_1^2]\varepsilon_1 +$$
$$\frac{2r_2^2}{r_2^2 - r_1^2}(1 + \frac{\nu_{01}}{1-2\nu_{01}}\frac{\psi_1(\lambda_1)}{\phi_1(\lambda_1)})\varepsilon_2 +$$

$$\frac{\nu_{01}}{1-2\nu_{01}}\frac{\psi_1(\lambda_1)}{\phi_1(\lambda_1)}\varepsilon_{zz} = \frac{1}{2G_{01}\phi_1(\lambda_1)}[\eta_{01}\eta_1(\lambda_1)\cdot\alpha_1\cdot\tilde{\lambda}_1 - p_a] \quad (2.50)$$

$$4G_{01}\phi_1(\lambda_1)\frac{r_1^2}{r_2^2 - r_1^2}(1 + \frac{\nu_{01}}{1-2\nu_{01}}\frac{\psi_1(\lambda_1)}{\phi_1(\lambda_1)})\varepsilon_1 -$$

$$\{\frac{2G_{02}\phi_2(\lambda_2)}{r_3^2 - r_2^2}[r_3^2 + (1 + \frac{2\nu_{02}}{1-2\nu_{02}}\frac{\psi_2(\lambda_2)}{\phi_2(\lambda_2)})\cdot r_2^2] +$$

$$\frac{2G_{01}\phi_1(\lambda_1)}{r_2^2 - r_1^2}[r_1^2 + (1 + \frac{2\nu_{01}}{1-2\nu_{01}}\frac{\psi_1(\lambda_1)}{\phi_1(\lambda_1)})\cdot r_2^2]\}\varepsilon_2 +$$

$$2[\frac{G_{02}\nu_{02}}{1-2\nu_{02}}\psi_2(\lambda_2) - \frac{G_{01}\nu_{01}}{1-2\nu_{01}}\psi_1(\lambda_1)]\cdot\varepsilon_{zz} =$$
$$\eta_{02}\eta_2(\lambda_2)\alpha_2\tilde{\lambda}_2 - \eta_{01}\eta_1(\lambda_1)\alpha_1\tilde{\lambda}_1 \quad (2.51)$$

$$\frac{1}{r_3^2 - r_2^2}[r_2^2 + (1 + \frac{2\nu_{02}}{1-2\nu_{02}}\frac{\psi_2(\lambda_2)}{\phi_2(\lambda_2)})r_3^2]\varepsilon_3 -$$
$$\frac{2r_2^2}{r_3^2 - r_2^2}[1 + \frac{\nu_{02}}{1-2\nu_{02}}\frac{\psi_2(\lambda_2)}{\phi_2(\lambda_2)}]\varepsilon_2 +$$

$$\frac{\nu_{02}}{1-2\nu_{02}}\frac{\psi_2(\lambda_2)}{\phi_2(\lambda_2)}\varepsilon_{zz}=\frac{1}{2G_{02}\phi_2(\lambda_2)}[\eta_{02}\eta_2(\lambda_2)\alpha_2\tilde{\lambda}_2-p_b] \quad (2.52)$$

$$4\sum_{n=1}^{2}G_{on}\psi_n(\lambda_n)\frac{\nu_{on}}{1-2\nu_{on}}(\varepsilon_{n+1}r_{n+1}^2-\varepsilon_n r_n^2)+$$

$$2\varepsilon_{zz}\sum_{n=1}^{2}G_{on}\phi_n(\lambda_n)(r_{n+1}^2-r_n^2)(1+\frac{\nu_{on}}{1-2\nu_{on}}\frac{\psi_n(\lambda_n)}{\phi_n(\lambda_n)})=$$

$$\frac{1}{\pi}P+\sum_{n=1}^{2}\eta_{on}\eta_n(\lambda_n)(r_{n+1}^2-r_n^2)\alpha_n\tilde{\lambda}_n \quad (2.53)$$

Therewith, we note the following: Because of heterogeneity of polymer materials of the first and second layers of the pipe, mathematical relation (2.34) in conformity to the two-layer pipe, that is, between the swelling parameter. λ_1 and λ_2 will be in the form:

$$\lambda_2=\frac{q_2}{q_1}\lambda_1 \quad (2.54)$$

where, q_1 and q_2 are the coefficients of linear dependences $\lambda=\lambda(t)$ for each layer of the sandwich pipe. They are determined experimentally. For example, accepting the value $\lambda_1=\lambda_{1\max}$ for the material offirst layer, then its appropriate value for the parameter λ_2 will be determined by relation (2.54) that is, in the form:

$$\lambda_2=\frac{q_2}{q_1}\lambda_{1\max}=\lambda_2^* \quad (2.56)$$

The system of equations (2.50)–(2.54) are linear-algebraic system of four equations with respect to ε_1, ε_2, ε_3, $\varepsilon_4=\varepsilon_z$. Write this system in the matrix form:

$$a_{ij}\varepsilon_j=b_i \quad (2.57)$$

where,

$$a_{11} = -\frac{1}{r_2^2 - r_1^2}[r_2^2 + (1 + \frac{2\nu_{01}}{1-2\nu_{01}}\frac{\psi_1(\lambda_1)}{\phi_1(\lambda_1)})r_1^2],$$

$$a_{12} = \frac{2r_2^2}{r_2^2 - r_1^2}(1 + \frac{\nu_{01}}{1-2\nu_{01}}\frac{\psi_1(\lambda_1)}{\phi_1(\lambda_1)})$$

$$a_{13} = 0,\ a_{14} = \frac{\nu_{01}}{1-2\nu_{01}}\frac{\psi_1(\lambda_1)}{\phi_1(\lambda_1)},$$

$$b_1 = \frac{1}{2G_0\phi_1(\lambda_1)}[\eta_{01}\eta_1(\lambda_1)\alpha_1\tilde{\lambda}_1 - p_a]$$

$$a_{21} = 4G_{01}\phi_1(\lambda_1)\frac{r_1^2}{r_2^2 - r_1^2}(1 + \frac{\nu_{01}}{1-2\nu_{01}}\frac{\psi_1(\lambda_1)}{\phi_1(\lambda_1)}),$$

$$-a_{22} = \frac{2G_{02}\phi_2(\lambda_2)}{r_3^2 - r_2^2}[r_3^2 + (1 + \frac{2\nu_{02}}{1-2\nu_{02}}\frac{\psi_2(\lambda_2)}{\phi_2(\lambda_2)})r_2^2] +$$

$$\frac{2G_{01}\phi_1(\lambda_1)}{r_2^2 - r_1^2}[r_1^2 + (1 + \frac{2\nu_{01}}{1-2\nu_{01}}\frac{\psi_1(\lambda_1)}{\phi_1(\lambda_1)})r_2^2]$$

$$a_{23} = 4G_{02}\phi_2(\lambda_2)\frac{r_3^2}{r_3^2 - r_2^2}(1 + \frac{\nu_{02}}{1-2\nu_{02}}\cdot\frac{\psi_2(\lambda_2)}{\phi_2(\lambda_2)}),$$

$$a_{24} = 2[\frac{G_{02}\nu_{02}}{1-2\nu_{02}}\psi_2(\lambda_2) - \frac{G_{01}\nu_{01}}{1-2\nu_{01}}\psi_1(\lambda_1)],$$

$$b_2 = \eta_{02}\eta_2(\lambda_2)\alpha_2\tilde{\lambda}_2 - \eta_{01}\eta_1(\lambda_1)\alpha_1\tilde{\lambda}_1,$$

$$a_{31} = 0,\ a_{32} = -\frac{2r_2^2}{r_3^2 - r_2^2}[1 + \frac{\nu_{02}}{1-2\nu_{02}}\frac{\psi_2(\lambda_2)}{\phi_2(\lambda_2)}],$$

$$a_{33} = \frac{1}{r_3^2 - r_2^2}[r_2^2 + (1 + \frac{2\nu_{02}}{1-2\nu_{02}}\frac{\psi_2(\lambda_2)}{\phi_2(\lambda_2)})r_3^2],$$

$$a_{34} = \frac{\nu_{02}}{1-2\nu_{02}}\frac{\psi_2(\lambda_2)}{\phi_2(\lambda_2)},$$

$$b_3 = \frac{1}{2G_{02}\phi_2(\lambda_2)}[\eta_{02}\eta_2(\lambda_2)\alpha_2\tilde{\lambda}_2 - p_b],$$

$$a_{41} = -4G_{01}\psi_1(\lambda_1)\frac{\nu_{01}}{1-2\nu_{01}}r_1^2,$$

$$a_{42} = 4r_2^2[G_{01}\psi_1(\lambda_1)\frac{\nu_{01}}{1-2\nu_{01}} - G_{02}\frac{\nu_{02}}{1-2\nu_{02}}\psi_2(\lambda_2)]$$

$$a_{43} = 4r_3^2 G_{02}\frac{\nu_{02}}{1-2\nu_{02}}\psi_2(\lambda_2),$$

$$a_{44} = 2\sum_{n=1}^{2} G_{0n}\phi_n(\lambda_n)(r_{n+1}^2 - r_n^2)[1 + \frac{\nu_{0n}}{1-2\nu_{0n}}\frac{\psi_n(\lambda_n)}{\phi_n(\lambda_n)}],$$

$$b_4 = \frac{1}{\pi}P + \sum_{n=1}^{2}\eta_{0n}\eta_n(\lambda_n)(r_{n+1}^2 - r_n^2)\alpha_n\tilde{\lambda}_n \tag{2.58}$$

By Kramer's law, deformations ε_{ij} will equal:

$$\varepsilon_j = \frac{\left|D_{x_j}\right|_4}{\left|a_{ij}\right|_4} = \frac{\sum_{i=1}^{4}(-1)^{j+i} b_i B_{ij}^{\varepsilon_j}}{\sum_{k=1}^{4}(-1)^{1+k} a_{1k} A_{1k}} \tag{2.59}$$

Here A_{1k} are the minors of the determinant $\left|a_{ij}\right|_4$, $B_{ij}^{\varepsilon_j}$ are the minors of the determinants $\left|D_{x_j}\right|_4$. In the expanded form, (2.59) will take the form:

$$\varepsilon_1 = \frac{\sum_{i=1}^{4}(-1)^{1+i} b_i B_{i1}^{\varepsilon_1}}{\sum_{k=1}^{4}(-1)^{1+k} a_{1k} A_{1k}}, \quad \varepsilon_2 = \frac{\sum_{i=1}^{4}(-1)^{2+i} b_i B_{i2}^{\varepsilon_2}}{\sum_{k=1}^{4}(-1)^{1+k} a_{1k} A_{1k}},$$

$$\varepsilon_3 = \frac{\sum_{i=1}^{4}(-1)^{3+i} b_i B_{i3}^{\varepsilon_3}}{\sum_{k=1}^{4}(-1)^{1+k} a_{1k} A_{1k}}, \quad \varepsilon_4 = \varepsilon_z = \frac{\sum_{i=1}^{4}(-1)^{4+i} b_i B_{i4}^{\varepsilon_4}}{\sum_{k=1}^{4}(-1)^{1+k} a_{1k} A_{1k}} \tag{2.60}$$

Being given mechanical and physico–chemical properties of layers of a two layer pipe E_{0i}, ν_{0i}, G_{0i}, λ_i, α_{0i}, $\tilde{\lambda}_{0i}$, $\varphi_i(\lambda_i)$, $\psi_i(\lambda_i)$, η_{0i}, $\eta_i(\lambda_i)$, and also geometrical characteristics of construction r_1, r_2, r_3, the deformation ε_1, ε_2, ε_3, $\varepsilon_4 = \varepsilon_z$ by formula (2.60) are defined uniquely depending on external loads p_a, p_b, P with regard to effect of changeability of physico–chemical properties of layers. Having defined the numerical values of tangential and longitudinal deformations ε_1, ε_2, ε_3, $\varepsilon_4 = \varepsilon_z$ of the first and second layers, by formula (2.17)–(2.18) we find stress distribution on the layers thickness in the form:

For first layer:

$$\frac{1}{2G_{01}\phi_1(\lambda_1)}\begin{pmatrix}\sigma_{yy}\\ \sigma_{rr}\end{pmatrix}=\frac{\nu_{01}}{1-2\nu_{01}}\frac{\psi_1(\lambda_1)}{\phi_1(\lambda_1)}\varepsilon_{zz}+[1+$$
$$\frac{2\nu_{01}}{1-2\nu_{01}}\cdot\frac{\psi_1(\lambda_1)}{\phi_1(\lambda_1)}]\frac{\varepsilon_2 r_2^2-\varepsilon_1 r_1^2}{r_2^2-r_1^2}\pm$$
$$\frac{1}{r^2}\cdot\frac{\varepsilon_2-\varepsilon_1}{r_2^{-2}-r_1^{-2}}-\frac{\eta_{o1}}{2G_{01}}\frac{\eta_1(\lambda_1)}{\phi_1(\lambda_1)}\alpha_1\tilde{\lambda}_1 \tag{2.61}$$

$$\frac{1}{2G_{01}\phi_1(\lambda_1)}\sigma_{zz}=[1+\frac{\nu_{01}}{1-2\nu_{01}}\cdot\frac{\psi_1(\lambda_1)}{\phi_1(\lambda_1)}]\varepsilon_{zz}+$$
$$\frac{2\nu_{01}}{1-2\nu_{01}}\frac{\psi_1(\lambda_1)}{\phi_1(\lambda_1)}\frac{\varepsilon_2 r_2^2-\varepsilon_1 r_1^2}{r_2^2-r_1^2}-\frac{\eta_{01}}{2G_{01}}\frac{\eta_1(\lambda_1)}{\phi_1(\lambda_1)}\cdot\alpha_1\cdot\tilde{\lambda}_1 \tag{2.62}$$

for second layer:

$$\frac{1}{2G_{02}\phi_2(\lambda_2)}\begin{pmatrix}\sigma_{yy}\\ \sigma_{rr}\end{pmatrix}=\frac{\nu_{02}}{1-2\nu_{02}}\frac{\psi_2(\lambda_2)}{\phi_2(\lambda_2)}\varepsilon_{zz}+[1+$$
$$\frac{2\nu_{02}}{1-2\nu_{02}}\cdot\frac{\psi_2(\lambda_2)}{\phi_2(\lambda_2)}]\frac{\varepsilon_3 r_3^2-\varepsilon_2 r_2^2}{r_3^2-r_2^2}\pm$$
$$\frac{1}{r^2}\cdot\frac{\varepsilon_3-\varepsilon_2}{r_3^{-2}-r_2^{-2}}-\frac{\eta_{02}}{2G_{02}}\frac{\eta_2(\lambda_2)}{\phi_2(\lambda_2)}\alpha_2\tilde{\lambda}_2 \tag{2.63}$$

$$\frac{1}{2G_{02}\varphi_2(\lambda_2)}\sigma_{zz}=[1+\frac{\nu_{02}}{1-2\nu_{02}}\cdot\frac{\psi_2(\lambda_2)}{\varphi_2(\lambda_2)}]\varepsilon_{zz}+$$
$$\frac{2\nu_{02}}{1-2\nu_{02}}\frac{\psi_2(\lambda_2)}{\varphi_2(\lambda_2)}\frac{\varepsilon_3 r_3^2-\varepsilon_2 r_2^2}{r_3^2-r_2^2}-\frac{\eta_{02}}{2G_{02}}\frac{\eta_2(\lambda_2)}{\varphi_2(\lambda_2)}\cdot\alpha_2\cdot\tilde{\lambda}_2 \tag{2.64}$$

For $r = r_1$, $r = r_2$, and $r = r_3$ by formulae (2.3.61)–(2.64), determine radial stresses σ_1^+, σ_2^+, σ_2^-, σ_3^- and in the same way, tangential stresses on the boundary surfaces, in the form:

$$\frac{1}{2G_{01}\phi_1(\lambda_1)}\sigma_1^+ = \frac{\nu_{01}}{1-2\nu_{01}}\frac{\psi_1(\lambda_1)}{\phi_1(\lambda_1)}\varepsilon_4 + \frac{2r_2^2}{r_2^2-r_1^2}[1+ \frac{\nu_{01}}{1-2\nu_{01}}\cdot\frac{\psi_1(\lambda_1)}{\phi_1(\lambda_1)}]\cdot\varepsilon_2 - \frac{1}{r_2^2-r_1^2}[r_2^2+(1+\frac{2\nu_{01}}{1-2\nu_{01}}\cdot \frac{\psi_1(\lambda_1)}{\phi_1(\lambda_1)})r_1^2]\cdot\varepsilon_1 - \frac{\eta_{01}}{2G_{01}}\frac{\eta_1(\lambda_1)}{\phi_1(\lambda_1)}\alpha_1\tilde{\lambda}_1 \quad (2.65)$$

$$\frac{1}{2G_{01}\phi_1(\lambda_1)}\sigma_2^- = \frac{\nu_{01}}{1-2\nu_{01}}\frac{\psi_1(\lambda_1)}{\phi_1(\lambda_1)}\varepsilon_4 + \frac{1}{r_2^2-r_1^2}[r_1^2+ (1+\frac{2\nu_{01}}{1-2\nu_{01}}\cdot\frac{\psi_1(\lambda_1)}{\phi_1(\lambda_1)})r_2^2]\cdot\varepsilon_2 - \frac{2r_1^2}{r_2^2-r_1^2}[1+ \frac{\nu_{01}}{1-2\nu_{01}}\cdot\frac{\psi_1(\lambda_1)}{\phi_1(\lambda_1)}]\cdot\varepsilon_1 - \frac{\eta_{01}}{2G_{01}}\frac{\eta_1(\lambda_1)}{\phi_1(\lambda_1)}\alpha_1\tilde{\lambda}_1 \quad (2.66)$$

$$\frac{1}{2G_{02}\phi_2(\lambda_2)}\sigma_2^+ = \frac{\nu_{02}}{1-2\nu_{02}}\frac{\psi_2(\lambda_2)}{\phi_2(\lambda_2)}\varepsilon_4 + \frac{2r_3^2}{r_3^2-r_2^2}[1+ \frac{\nu_{02}}{1-2\nu_{02}}\cdot\frac{\psi_2(\lambda_2)}{\phi_2(\lambda_2)}]\cdot\varepsilon_3 - \frac{1}{r_3^2-r_2^2}[r_3^2+(1+\frac{2\nu_{02}}{1-2\nu_{02}}\cdot \frac{\psi_2(\lambda_2)}{\phi_2(\lambda_2)})r_2^2]\cdot\varepsilon_2 - \frac{\eta_{02}}{2G_{02}}\frac{\eta_2(\lambda_2)}{\phi_2(\lambda_2)}\alpha_2\tilde{\lambda}_2 \quad (2.67)$$

$$\frac{1}{2G_{02}\phi_2(\lambda_2)}\sigma_3^- = \frac{\nu_{02}}{1-2\nu_{02}}\frac{\psi_2(\lambda_2)}{\phi_2(\lambda_2)}\varepsilon_4 + \frac{1}{r_3^2 - r_2^2}[r_2^2 +$$
$$(1+\frac{2\nu_{02}}{1-2\nu_{02}}\cdot\frac{\psi_2(\lambda_2)}{\phi_2(\lambda_2)})r_3^2]\cdot\varepsilon_3 - \frac{2r_2^2}{r_3^2-r_2^2}[1+ \qquad (2.68)$$
$$\frac{\nu_{02}}{1-2\nu_{02}}\cdot\frac{\psi_2(\lambda_2)}{\phi_2(\lambda_2)}]\cdot\varepsilon_2 - \frac{\eta_{02}}{2G_{02}}\frac{\eta_2(\lambda_2)}{\phi_2(\lambda_2)}\alpha_2\tilde{\lambda}_2$$

Special case: Consider a two-layer pipe consisting of polymer material situated only under the action of corrosive liquid medium. This case is interesting in the sense of studying and knowing quality and quantity character of stress and deformation distribution in the structure's layer at the expense of influence of physico–chemical changes of the layer's material. Therewith, determination of maximal breaking off stresses on the boundaries between layers depending on change of physico–chemical properties of the layers material, and arising in the layers of thrust mass forces is of special interest. For solving this problem, we use formula (2.65)–(2.68), where we equate the external mechanical parameters P, p_a, and p_b to zero. We conduct numerical calculations under the following input data. The internal layer of a two-layer pipe is made up of a polymer material of the brand "polyethylene of lower density" $(PELD)$ and the external layer made up of polymer of brand "polystyrene shock-resistant" (PS) with the following geometrical sizes:

- Geometrical sizes of the construction of two-layer pipe:

$$r_1 = 0.1\text{m}, \quad r_2 = 0.13\text{m}, \quad r_3 = 0.15\text{m}$$

- Mechanical properties and physico–chemical characteristics of the internal layer (of PELD):

$\alpha_1 = 0.3452$, $\lambda_{1\max} = 0.0289$, $E_{01} = 34MPa$, $\nu_{01} = 0.4$, $E_{0\min} = 21MPa$,

$\nu_{1\max} = 0.435$, $b_1 = 0.61765$, $k_1 = 0.035$, $q_1 = 0.0104$, $\tilde{\alpha}_1 = \alpha_1\lambda_{1\max} = 0.01$,

$2G_{01} = 24.2857MPa$, $a_{01} = 48,5714$, $\eta_{01} = 170$, $\phi_1(\lambda_1)\big|_{\tilde{\lambda}_1=1} = 0.6026$,

$\psi_1(\lambda_1)\big|_{\tilde{\lambda}_1=1} = 1.0082$, $\eta_1(\lambda_1)\big|_{\tilde{\lambda}_1=1} = 0.0274$, $\gamma_1 = 9025.2\dfrac{N}{m^3}$

- Mechanical properties and physico–chemical characteristics of the external layer (of PS):

$\alpha_2 = 0.2333$, $\lambda_{2\max} = 0.02$, $E_{02} = 153.5MPa$, $\nu_{02} = 0.33$, $\nu_{2\max} = 0.37$

$b_2 = 0.8795$, $k_2 = 0.04$, $q_2 = 0.0072$, $\tilde{\alpha}_2 = \alpha_2\lambda_{2\max} = 0.0047$,

$2G_{02} = 115.4135MPa$, $a_{02} = 112.019$, $\eta_{02} = 451.4706$, $\phi_2(\lambda_2)\big|_{\tilde{\lambda}_2=1} = 0.8538$,

$\psi_2(\lambda_2)\big|_{\tilde{\lambda}_2=1} = 1.2518$, $\eta_2(\lambda_2)\big|_{\tilde{\lambda}_2=1} = 0.023$, $\gamma_2 = 10,300.5\dfrac{N}{m^3}$

In this case, algebraic system (2.57) with respect to contour tangential deformations ε_1, ε_2, ε_3 and longitudinal deformation $\varepsilon_4 = \varepsilon_{zz}$ will be in form:

$$\begin{cases} -13.5977\cdot\varepsilon_1 + 21.29\cdot\varepsilon_2 + 0\cdot\varepsilon_3 + 3.3462\cdot\varepsilon_4 = 0.11 - 0.0683\cdot p_a \\ 184.3615\cdot\varepsilon_1 - 1836.6211\cdot\varepsilon_2 + 1918.6245\cdot\varepsilon_3 + 91.2574\cdot\varepsilon_4 = 0.8146 \\ 0\cdot\varepsilon_1 - 14.6245\cdot\varepsilon_2 + 18.4705\cdot\varepsilon_3 + 1.423\cdot\varepsilon_4 = 0.0246 \\ -0.9794\cdot\varepsilon_1 - 3.0845\cdot\varepsilon_2 + 6.3102\cdot\varepsilon_3 + 1.776\cdot\varepsilon_4 = 0.0247 + 0.3185\cdot P \end{cases} \tag{2.69}$$

For $P = p_a = p_b = 0$, the solution of system (2.69) will be:

$$\varepsilon_1 = 0.0034,\ \varepsilon_2 = 0.0064,\ \varepsilon_3 = 0.006,\ \varepsilon_4 = 0.0057 \tag{2.70}$$

In this case, numerical values of displacements of the points of boundary surfaces will equal:

$$w\big|_{r_1=0,1m} = \varepsilon_1 r\big|_{r_1=0,1m} = 0.34mm,$$

$$w\big|_{r_2=0.13\text{m}} = \varepsilon_2 r\big|_{r_2=0.13\text{m}} = 0.832\text{mm},$$

$$w\big|_{r_3=0.15\text{m}} = \varepsilon_1 r\big|_{r_3=0.15\text{m}} = 0.9\text{mm} \tag{2.71}$$

Substituting the numerical values of tangential and longitudinal deformations ε_1, ε_2, ε_3, $\varepsilon_4 = \varepsilon_z$ (2.70), from formulae (2.61)–(2.64) we find stress distribution on the thickness of the first and second layers. They will be in the form:

For first layer:

$$\begin{pmatrix} \sigma_y \\ \sigma_r \end{pmatrix} = 0.4654 \pm 0.0001 \cdot \frac{1}{r^2}$$
$$\sigma_{zz} = 0.1741 \qquad \text{for } 0.1\text{m} \le r_1 \le 0.13\text{m} \tag{2.72}$$

For second layer:

$$\begin{pmatrix} \sigma_y \\ \sigma_r \end{pmatrix} = -0.2759 \pm 0.003 \cdot \frac{1}{r^2} \qquad \text{for } 0.13\text{m} \le r_1 \le 0.15\text{m} \tag{2.73}$$
$$\sigma_{zz} = -0.0591$$

From formulae (2.72) and (2.73) the numerical values of radial stresses at the points of boundary surfaces are determined:

$$\sigma_1^+ = \sigma_r\big|_{r_1=0.1m} = 0.4754MPa$$

$$\sigma_2^- = \sigma_r\big|_{r_2=0.13m} = 0.4603MPa$$

$$\sigma_2^+ = \sigma_r\big|_{r_2=0.13m} = -0.4534MPa$$

$$\sigma_3^- = \sigma_r\Big|_{r_3=0.15m} = -0.4092MPa \tag{2.74}$$

From (2.74), it is seen that radial stresses at points of boundary surfaces of a two-layer pipe situated only under the action of corrosive liquid medium, attain the order $4.7\frac{\text{kgf}}{\text{cm}^2}$. In other words, this is a result of arising thrust mass force from physico–chemical change of polymer layers of pipe.

Summarizing the above mentioned ones, we can say that formulae (2.70)–(2.74) completely define the numerical values of displacements, deformations, and also stresses both on the thickness of pipe's layers and at the contact points depending on physico–chemical changeability of structure's layers.

2.4 STRESS–STRAIN STATE OF A THREE-LAYER PIPE SITUATED UNDER THE ACTION OF CORROSIVE LIQUID MEDIUM AND EXTERNAL LOADS

In this section (Aliyev, 2012), we shall investigate stress–strain state of a three-layer pipe consisting of polymer materials with regard to change of physico–chemical properties of layers that arise after coming in contact with corrosive medium (Figure 2.4).

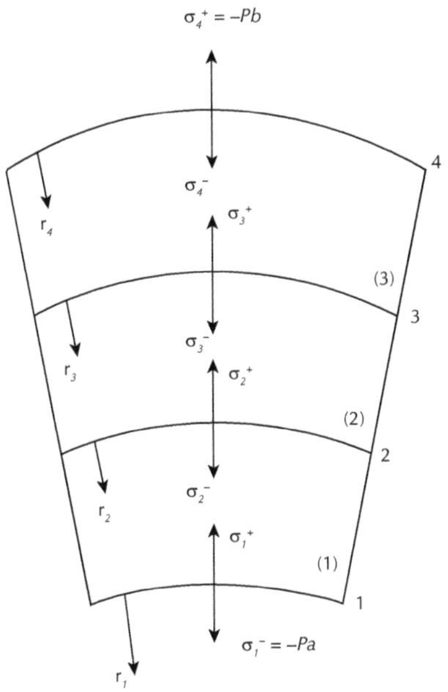

FIGURE 2.4 Three-layer pipe made up of polymer material

Moreover, it is assumed that the pipe's layers are piecewise homogeneous with the following mechanical and physico–chemical properties:

Layer I—E_{01}, ν_{01}, G_{01}, η_{01}, λ_1, α_1, $\phi_1(\lambda_1)$, $\psi_1(\lambda_1)$, and $\eta_1(\lambda_1)$

Layer II—E_{02}, ν_{02}, G_{02}, η_{02}, λ_2, α_2, $\phi_2(\lambda_2)$, $\psi_2(\lambda_2)$, and $\eta_2(\lambda_2)$

Layer III—E_{03}, ν_{03}, G_{03}, η_{03}, λ_3, α_3, $\phi_3(\lambda_3)$, $\psi_3(\lambda_3)$, and $\eta_3(\lambda_3)$ (2.75)

Quantities of stresses, deformations in layers and breaking off stresses between the layers defining adhesive strength of a sandwich pipe situated under the joint action of external loads p_a, p_b, P and swelling forces of layers λ_n that arise when it contacts with corrosive liquid media, were determined. The problem solution will be determined by means of equations obtained earlier. For that in the system of equations (2.23), (2.30), (2.31), and (2.33), we accept $n=3$. The problem is reduced to the system of five algebraic equations with respect to unknown values of contour values of tangential ε_1, ε_2, ε_3, ε_4 and longitudinal $\varepsilon_5 = \varepsilon_{zz}$ deformations in form:

$$-\frac{1}{r_2^2 - r_1^2}[r_2^2 + (1 + \frac{2\nu_{01}}{1-2\nu_{01}}\frac{\psi_1(\lambda_1)}{\phi_1(\lambda_1)})r_1^2]\varepsilon_1 + \frac{2r_2^2}{r_2^2 - r_1^2}(1 +$$

$$\frac{\nu_{01}}{1-2\nu_{01}}\frac{\psi_1(\lambda_1)}{\phi_1(\lambda_1)})\varepsilon_2 + \frac{\nu_{01}}{1-2\nu_{01}}\frac{\psi_1(\lambda_1)}{\phi_1(\lambda_1)}\varepsilon_5$$

$$= \frac{1}{2G_{01}\phi_1(\lambda_1)}[\eta_{01}\eta_1(\lambda_1)\cdot\alpha_1\cdot\tilde{\lambda}_1 - p_a] \quad (2.76)$$

$$4G_{01}\phi_1(\lambda_1)\frac{r_1^2}{r_2^2 - r_1^2}(1 + \frac{\nu_{01}}{1-2\nu_{01}}\frac{\psi_1(\lambda_1)}{\phi_1(\lambda_1)})\varepsilon_1 -$$

$$\{\frac{2G_{02}\phi_2(\lambda_2)}{r_3^2 - r_2^2}[r_3^2 + (1 + \frac{2\nu_{02}}{1-2\nu_{02}}\frac{\psi_2(\lambda_2)}{\phi_2(\lambda_2)})\cdot r_2^2]$$

$$+\frac{2G_{01}\phi_1(\lambda_1)}{r_2^2 - r_1^2}[r_1^2 + (1 + \frac{2\nu_{01}}{1-2\nu_{01}}\frac{\psi_1(\lambda_1)}{\phi_1(\lambda_1)})\cdot r_2^2]\}\varepsilon_2$$

$$+4G_{02}\phi_2(\lambda_2)\frac{r_3^2}{r_3^2 - r_2^2}(1 + \frac{\nu_{02}}{1-2\nu_{02}}\frac{\psi_2(\lambda_2)}{\phi_2(\lambda_2)})\varepsilon_3$$

$$+2[\frac{G_{02}\nu_{02}}{1-2\nu_{02}}\psi_2(\lambda_2) - \frac{G_{01}\nu_{01}}{1-2\nu_{01}}\psi_1(\lambda_1)]\cdot\varepsilon_5$$

$$= \eta_{02}\eta_2(\lambda_2)\alpha_2\tilde{\lambda}_2 - \eta_{01}\eta_1(\lambda_1)\alpha_1\tilde{\lambda}_1 \quad (2.77)$$

$$4G_{02}\phi_2(\lambda_2)\frac{r_2^2}{r_3^2-r_2^2}[1+\frac{\nu_{02}}{1-2\nu_{02}}\frac{\psi_2(\lambda_2)}{\phi_2(\lambda_2)}]\varepsilon_2$$

$$-\{\frac{2G_{03}\phi_3(\lambda_3)}{r_4^2-r_3^2}[r_4^2+(1+\frac{2\nu_{03}}{1-2\nu_{03}}\frac{\psi_3(\lambda_3)}{\phi_3(\lambda_3)})r_3^2]$$

$$+\frac{2G_{02}\phi(\lambda_2)}{r_3^2-r_2^2}\left[r_2^2+(1+\frac{2\nu_{02}}{1-2\nu_{02}}\frac{\psi_2(\lambda_2)}{\phi_2(\lambda_2)})\cdot r_3^2]\right\}\varepsilon_3+$$

$$4G_{03}\phi_3(\lambda_3)\frac{r_4^2}{r_4^2-r_3^2}(1+\frac{\nu_{03}}{1-2\nu_{03}}\frac{\psi_3(\lambda_3)}{\phi_3(\lambda_3)})\varepsilon_4+$$

$$+2[\frac{G_{03}\nu_{03}}{1-2\nu_{03}}\psi_3(\lambda_3)-\frac{G_{02}\nu_{02}}{1-2\nu_{02}}\psi_2(\lambda_2)]\varepsilon_5$$

$$=\eta_{03}\eta_3(\lambda_3)\alpha_3\tilde{\lambda}_3-\eta_{02}\eta_2(\lambda_2)\alpha_2\tilde{\lambda}_2 \tag{2.78}$$

$$\frac{1}{r_4^2-r_3^2}[r_3^2+(1+\frac{2\nu_{03}}{1-2\nu_{03}}\frac{\psi_3(\lambda_3)}{\phi_3(\lambda_3)})r_4^2]\varepsilon_4$$

$$-\frac{2r_3^2}{r_4^2-r_3^2}\cdot[1+\frac{\nu_{03}}{1-2\nu_{03}}\frac{\psi_3(\lambda_3)}{\psi_3(\lambda_3)}]\varepsilon_3+\frac{\nu_{03}}{1-2\nu_{03}}\frac{\psi_3(\lambda_3)}{\phi_3(\lambda_3)}\varepsilon_5$$

$$=\frac{1}{2G_{03}\phi_3(\lambda_3)}[\eta_{03}\eta_3(\lambda_3)\alpha_3\tilde{\lambda}_3-p_b] \tag{2.79}$$

$$4\sum_{n=1}^{3}G_{on}\psi_n(\lambda_n)\frac{\nu_{on}}{1-2\nu_{on}}(\varepsilon_{n+1}r_{n+1}^2-\varepsilon_n r_n^2)+$$

$$2\varepsilon_5\sum_{n=1}^{3}G_{on}\phi_n(\lambda_n)(r_{n+1}^2-r_n^2)(1+\frac{\nu_{on}}{1-2\nu_{on}}\frac{\psi_n(\lambda_n)}{\phi_n(\lambda_n)})$$

$$=\frac{1}{\pi}P+\sum_{n=1}^{3}\eta_{on}\eta_n(\lambda_n)(r_{n+1}^2-r_n^2)\alpha_n\tilde{\lambda}_n \tag{2.80}$$

In this problem, from the heterogeneity of the materials of the first, second, and third layers, mathematical relation between their swelling parameters λ_1, λ_2 and λ_3 by (2.34), is written in the form:

$$\lambda_n = \frac{q_n}{q_{n-1}} \lambda_{n-1} \tag{2.81}$$

In the expanded form:

$$\lambda_2 = \frac{q_2}{q_1} \lambda_1 , \ \lambda_3 = \frac{q_3}{q_2} \lambda_2 = \frac{q_3}{q_1} \lambda_1 \tag{2.82}$$

Thus, being given one of the values of the swelling parameter, the another two parameters are defined from formula (2.81).

Thus, the system of equations is a linear algebraic system of five equations with respect to ε_1, ε_2, ε_3, ε_4, $\varepsilon_5 = \varepsilon_{zz}$. Write this system in the matrix form:

$$a_{ij} \varepsilon_j = b_i \tag{2.83}$$

where,

$$a_{11} = -\frac{1}{r_2^2 - r_1^2} [r_2^2 + (1 + \frac{2\nu_{01}}{1 - 2\nu_{01}} \frac{\psi_1(\lambda_1)}{\phi_1(\lambda_1)}) r_1^2],$$

$$a_{12} = \frac{2r_2^2}{r_2^2 - r_1^2} (1 + \frac{\nu_{01}}{1 - 2\nu_{01}} \frac{\psi_1(\lambda_1)}{\phi_1(\lambda_1)})$$

$$a_{13} = a_{14} = 0 , \ a_{15} = \frac{\nu_{01}}{1 - 2\nu_{01}} \frac{\psi_1(\lambda_1)}{\phi_1(\lambda_1)},$$

$$b_1 = \frac{1}{2G_{01}\phi_1(\lambda_1)} [\eta_{01}\eta_1(\lambda_1)\alpha_1\tilde{\lambda}_1 - p_a]$$

$$a_{21} = 4G_{01}\phi_1(\lambda_1)\frac{r_1^2}{r_2^2 - r_1^2}(1 + \frac{\nu_{01}}{1-2\nu_{01}}\frac{\psi_1(\lambda_1)}{\phi_1(\lambda_1)}),$$

$$-a_{22} = \frac{2G_{02}\phi_2(\lambda_2)}{r_3^2 - r_2^2}[r_3^2 + (1 + \frac{2\nu_{02}}{1-2\nu_{02}}\frac{\psi_2(\lambda_2)}{\phi_2(\lambda_2)})r_2^2]$$

$$+\frac{2G_{01}\phi_1(\lambda_1)}{r_2^2 - r_1^2}[r_1^2 + (1 + \frac{2\nu_{01}}{1-2\nu_{01}}\frac{\psi_1(\lambda_1)}{\phi_1(\lambda_1)})r_2^2]$$

$$a_{23} = 4G_{02}\phi_2(\lambda_2)\frac{r_3^2}{r_3^2 - r_2^2}(1 + \frac{\nu_{02}}{1-2\nu_{02}}\frac{\psi_2(\lambda_2)}{\phi_2(\lambda_2)}),\ \alpha_{24} = 0,$$

$$a_{25} = 2[\frac{G_{02}\nu_{02}}{1-2\nu_{02}}\psi_2(\lambda_2) - \frac{G_{01}\nu_{01}}{1-2\nu_{01}}\psi_1(\lambda_1)],$$

$$b_2 = \eta_{02}\eta_2(\lambda_2)\alpha_2\tilde{\lambda}_2 - \eta_{01}\eta_1(\lambda_1)\alpha_1\tilde{\lambda}_1,$$

$$a_{31} = 0,$$

$$a_{32} = 4G_{02}\phi_2(\lambda_2)\frac{r_2^2}{r_3^2 - r_2^2}[1 + \frac{\nu_{02}}{1-2\nu_{02}}\frac{\psi_2(\lambda_2)}{\phi_2(\lambda_2)}],$$

$$-a_{33} = \frac{2G_{03}\phi_3(\lambda_3)}{r_4^2 - r_3^2}[r_4^2 + (1 + \frac{2\nu_{03}}{1-2\nu_{03}}\frac{\psi_3(\lambda_3)}{\phi_3(\lambda_3)})r_3^2] +$$

$$\frac{2G_{02}\phi_2(\lambda_2)}{r_3^2 - r_2^2}[r_2^2 + (1 + \frac{2\nu_{02}}{1-2\nu_{02}}\frac{\psi_2(\lambda_2)}{\phi_2(\lambda_2)})r_3^2],$$

$$a_{34} = 4G_{03}\phi_3(\lambda_3)\frac{r_4^2}{r_4^2 - r_3^2}(1 + \frac{\nu_{03}}{1-2\nu_{03}}\frac{\psi_3(\lambda_3)}{\phi_3(\lambda_3)}),$$

$$a_{35} = 2[\frac{G_{03}\nu_{03}}{1-2\nu_{03}}\psi_3(\lambda_3) - \frac{G_{02}\nu_{02}}{1-2\nu_{02}}\psi_2(\lambda_2)],$$

$$b_3 = \eta_{03}\eta_3(\lambda_3)\alpha_3\tilde{\lambda}_3 - \eta_{02}\eta_2(\lambda_2)\alpha_2\tilde{\lambda}_2,$$

$$a_{41} = a_{42} = 0,$$

$$a_{43} = -\frac{2r_3^2}{r_4^2 - r_3^2}[1 + \frac{\nu_{03}}{1-2\nu_{03}}\frac{\psi_3(\lambda_3)}{\phi_3(\lambda_3)}],$$

$$a_{44} = \frac{1}{r_4^2 - r_3^2}[r_3^2 + (1 + \frac{2\nu_{03}}{1-2\nu_{03}}\frac{\psi_3(\lambda_3)}{\phi_3(\lambda_3)})r_4^2],$$

$$a_{45} = \frac{\nu_{03}}{1-2\nu_{03}}\frac{\psi_3(\lambda_3)}{\phi_3(\lambda_3)},$$

$$b_4 = \frac{1}{2G_{03}\phi_3(\lambda_3)}[\eta_{03}\eta_3(\lambda_3)\alpha_3\tilde{\lambda}_3 - p_b],$$

$$a_{51} = -4G_{01}\psi_1(\lambda_1)\frac{\nu_{01}}{1-2\nu_{01}}r_1^2,$$

$$a_{52} = 4r_2^2[G_{01}\psi_1(\lambda_1)\frac{\nu_{01}}{1-2\nu_{01}} - G_{02}\frac{\nu_{02}}{1-2\nu_{02}}\psi_2(\lambda_2)]$$

$$a_{53} = 4r_3^2[G_{02}\frac{\nu_{02}}{1-2\nu_{02}}\psi_2(\lambda_2) - G_{03}\frac{\nu_{03}}{1-2\nu_{03}}\psi_3(\lambda_3)],$$

$$a_{54} = 4r_4^2 G_{03}\frac{\nu_{03}}{1-2\nu_{03}}\psi_3(\lambda_3),$$

$$a_{55} = 2\sum_{n=1}^{3} G_{0n}\phi_n(\lambda_n)(r_{n+1}^2 - r_n^2)[1 + \frac{\nu_{0n}}{1-2\nu_{0n}}\frac{\psi_n(\lambda_n)}{\phi_n(\lambda_n)}],$$

$$b_5 = \frac{1}{\pi}P + \sum_{n=1}^{3}\eta_{0n}\eta_n(\lambda_n)(r_{n+1}^2 - r_n^2)\alpha_n\tilde{\lambda}_n \tag{2.84}$$

Solving linear-algebraic system (2.83), we determine tangential deformations ε_1, ε_2, ε_3, ε_4 and longitudinal deformation ε_5 on boundary surfaces of a sandwich pipe, in the form:

$$\varepsilon_j = \frac{\left|D_{x_j}\right|_5}{\left|a_{ij}\right|_5} = \frac{\sum_{i=1}^{5}(-1)^{j+i}b_i B_{ij}^{\varepsilon_j}}{\sum_{k=1}^{5}(-1)^{1+k}a_{1k}A_{1k}} \tag{2.85}$$

Here A_{1k} are the minors of the determinant $\left|a_{ij}\right|_5$, $B_{ij}^{\varepsilon_j}$ are the minors of determinants $\left|D_{x_j}\right|_5$. In expanded form, (2.85) will take the form:

$$\varepsilon_1 = \frac{\sum_{i=1}^{5} (-1)^{1+i} b_i B_{i1}^{\varepsilon_1}}{\sum_{k=1}^{5} (-1)^{1+k} a_{1k} A_{1k}}, \quad \varepsilon_2 = \frac{\sum_{i=1}^{5} (-1)^{2+i} b_i B_{i2}^{\varepsilon_2}}{\sum_{k=1}^{5} (-1)^{1+k} a_{1k} A_{1k}}$$

$$\varepsilon_3 = \frac{\sum_{i=1}^{5} (-1)^{3+i} b_i B_{i3}^{\varepsilon_3}}{\sum_{k=1}^{5} (-1)^{1+k} a_{1k} A_{1k}}, \quad \varepsilon_4 = \frac{\sum_{i=1}^{5} (-1)^{4+i} b_i B_{i4}^{\varepsilon_4}}{\sum_{k=1}^{5} (-1)^{1+k} a_{1k} A_{1k}},$$

$$\varepsilon_5 = \varepsilon_z = \frac{\sum_{i=1}^{5} (-1)^{5+i} b_i B_{i5}^{\varepsilon_5}}{\sum_{k=1}^{5} (-1)^{1+k} a_{1k} A_{1k}} \tag{2.86}$$

Being given mechanical and physico–chemical properties of the layers of three layer pipe E_{0i}, ν_{0i}, G_{0i}, λ_i, α_{0i}, $\tilde{\lambda}_{0i}$, $\varphi_i(\lambda_i)$, $\psi_i(\lambda_i)$, η_{0i}, $\eta_i(\lambda_i)$, and also geometrical characters of the construction r_1, r_2, r_3, the deformations ε_1, ε_2, ε_3, ε_4, $\varepsilon_5 = \varepsilon_z$, by formula (2.86) are uniquely determined depending on external loads p_a, p_b, and P with regard to the changeability of physico–chemical properties of layers. Having determined from system (2.86), the numerical values of tangential and longitudinal deformations ε_1, ε_2, ε_3, ε_4, $\varepsilon_5 = \varepsilon_z$ on the boundary surfaces of the first, second, and third layers, from formula (2.17) and (2.18), we find dependence of stress distribution on thickness of each layer on boundary values of tangential and longitudinal deformations in the form:

For first layer:

$$\frac{1}{2G_{01}\phi_1(\lambda_1)}\begin{pmatrix}\sigma_{yy}\\ \sigma_{rr}\end{pmatrix}=\frac{\nu_{01}}{1-2\nu_{01}}\frac{\psi_1(\lambda_1)}{\phi_1(\lambda_1)}\varepsilon_5+[1$$
$$+\frac{2\nu_{01}}{1-2\nu_{01}}\cdot\frac{\psi_1(\lambda_1)}{\phi_1(\lambda_1)}]\frac{\varepsilon_2 r_2^2-\varepsilon_1 r_1^2}{r_2^2-r_1^2}$$

$$\pm\frac{1}{r^2}\cdot\frac{\varepsilon_2-\varepsilon_1}{r_2^{-2}-r_1^{-2}}-\frac{\eta_{o1}}{2G_{01}}\frac{\eta_1(\lambda_1)}{\phi_1(\lambda_1)}\alpha_1\tilde{\lambda}_1 \tag{2.87}$$

$$\frac{1}{2G_{01}\phi_1(\lambda_1)}\sigma_{zz}=[1+\frac{\nu_{01}}{1-2\nu_{01}}\cdot\frac{\psi_1(\lambda_1)}{\phi_1(\lambda_1)}]\varepsilon_5+$$
$$\frac{2\nu_{01}}{1-2\nu_{01}}\frac{\psi_1(\lambda_1)}{\phi_1(\lambda_1)}\frac{\varepsilon_2 r_2^2-\varepsilon_1 r_1^2}{r_2^2-r_1^2}-\frac{\eta_{01}}{2G_{01}}\frac{\eta_1(\lambda_1)}{\phi_1(\lambda_1)}\cdot\alpha_1\cdot\tilde{\lambda}_1 \tag{2.88}$$

for the second layer:

$$\frac{1}{2G_{02}\phi_2(\lambda_2)}\begin{pmatrix}\sigma_{yy}\\ \sigma_{rr}\end{pmatrix}=\frac{\nu_{02}}{1-2\nu_{02}}\frac{\psi_2(\lambda_2)}{\phi_2(\lambda_2)}\varepsilon_5+[1+$$
$$\frac{2\nu_{02}}{1-2\nu_{02}}\cdot\frac{\psi_2(\lambda_2)}{\phi_2(\lambda_2)}]\frac{\varepsilon_3 r_3^2-\varepsilon_2 r_2^2}{r_3^2-r_2^2}$$

$$\pm\frac{1}{r^2}\cdot\frac{\varepsilon_3-\varepsilon_2}{r_3^{-2}-r_2^{-2}}-\frac{\eta_{02}}{2G_{02}}\frac{\eta_2(\lambda_2)}{\phi_2(\lambda_2)}\alpha_2\tilde{\lambda}_2 \tag{2.89}$$

$$\frac{1}{2G_{02}\phi_2(\lambda_2)}\sigma_{zz}=[1+\frac{\nu_{02}}{1-2\nu_{02}}\cdot\frac{\psi_2(\lambda_2)}{\phi_2(\lambda_2)}]\varepsilon_5+$$
$$\frac{2\nu_{02}}{1-2\nu_{02}}\frac{\psi_2(\lambda_2)}{\phi_2(\lambda_2)}\frac{\varepsilon_3 r_3^2-\varepsilon_2 r_2^2}{r_3^2-r_2^2}-\frac{\eta_{02}}{2G_{02}}\frac{\eta_2(\lambda_2)}{\phi_2(\lambda_2)}\cdot\alpha_2\cdot\tilde{\lambda}_2 \tag{2.90}$$

for the third layer:

$$\frac{1}{2G_{03}\phi_3(\lambda_3)}\begin{pmatrix}\sigma_{yy}\\ \sigma_{rr}\end{pmatrix}=\frac{\nu_{03}}{1-2\nu_{03}}\frac{\psi_3(\lambda_3)}{\phi_3(\lambda_3)}\varepsilon_5+[1+$$

$$\frac{2\nu_{03}}{1-2\nu_{03}}\cdot\frac{\psi_3(\lambda_3)}{\phi_3(\lambda_3)}]\frac{\varepsilon_4 r_4^2-\varepsilon_3 r_3^2}{r_4^2-r_3^2}\pm$$

$$\frac{1}{r^2}\cdot\frac{\varepsilon_4-\varepsilon_3}{r_4^{-2}-r_3^{-2}}-\frac{\eta_{03}}{2G_{03}}\frac{\eta_3(\lambda_3)}{\phi_3(\lambda_3)}\alpha_3\tilde{\lambda}_3 \tag{2.91}$$

$$\frac{1}{2G_{03}\phi_3(\lambda_3)}\sigma_{zz}=[1+\frac{\nu_{03}}{1-2\nu_{03}}\cdot\frac{\psi_3(\lambda_3)}{\phi_3(\lambda_3)}]\varepsilon_5+$$
$$\frac{2\nu_{03}}{1-2\nu_{03}}\frac{\psi_3(\lambda_3)}{\phi_3(\lambda_3)}\frac{\varepsilon_4 r_4^2-\varepsilon_3 r_3^2}{r_4^2-r_3^2}-\frac{\eta_{03}}{2G_{03}}\frac{\eta_3(\lambda_3)}{\phi_3(\lambda_3)}\cdot\alpha_3\cdot\tilde{\lambda}_3 \tag{2.92}$$

For $r=r_1$, $r=r_2$, $r=r_3$, and $r=r_4$ by formula (2.87)–(2.92), we define radial stresses σ_1^+, σ_2^+, σ_2^-, and σ_3^-, tangential stresses on boundary surfaces, in the form:

$$\frac{1}{2G_{01}\phi_1(\lambda_1)}\sigma_1^+=\frac{\nu_{01}}{1-2\nu_{01}}\frac{\psi_1(\lambda_1)}{\phi_1(\lambda_1)}\varepsilon_5+$$
$$\frac{2r_2^2}{r_2^2-r_1^2}[1+\frac{\nu_{01}}{1-2\nu_{01}}\cdot\frac{\psi_1(\lambda_1)}{\phi_1(\lambda_1)}]\cdot\varepsilon_2-$$
$$\frac{1}{r_2^2-r_1^2}[r_2^2+(1+\frac{2\nu_{01}}{1-2\nu_{01}}\cdot\frac{\psi_1(\lambda_1)}{\phi_1(\lambda_1)})r_1^2]\cdot\varepsilon_1-$$
$$\frac{\eta_{01}}{2G_{01}}\frac{\eta_1(\lambda_1)}{\phi_1(\lambda_1)}\alpha_1\tilde{\lambda}_1 \tag{2.93}$$

$$\frac{1}{2G_{01}\phi_1(\lambda_1)}\sigma_2^-=\frac{\nu_{01}}{1-2\nu_{01}}\frac{\psi_1(\lambda_1)}{\phi_1(\lambda_1)}\varepsilon_5$$
$$+\frac{1}{r_2^2-r_1^2}[r_1^2+(1+\frac{2\nu_{01}}{1-2\nu_{01}}\cdot\frac{\psi_1(\lambda_1)}{\phi_1(\lambda_1)})r_2^2]\cdot\varepsilon_2 \tag{2.94}$$
$$-\frac{2r_1^2}{r_2^2-r_1^2}[1+\frac{\nu_{01}}{1-2\nu_{01}}\cdot\frac{\psi_1(\lambda_1)}{\phi_1(\lambda_1)}]\cdot\varepsilon_1$$
$$-\frac{\eta_{01}}{2G_{01}}\frac{\eta_1(\lambda_1)}{\phi_1(\lambda_1)}\alpha_1\tilde{\lambda}_1$$

$$\begin{aligned}\frac{1}{2G_{02}\phi_2(\lambda_2)}\sigma_2^+ &= \frac{\nu_{02}}{1-2\nu_{02}}\frac{\psi_2(\lambda_2)}{\phi_2(\lambda_2)}\varepsilon_5 \\ &+\frac{2r_3^2}{r_3^2-r_2^2}[1+\frac{2\nu_{02}}{1-2\nu_{02}}\cdot\frac{\psi_2(\lambda_2)}{\phi_2(\lambda_2)}]\cdot\varepsilon_3 \\ &-\frac{1}{r_3^2-r_2^2}[r_3^2+(1+\frac{2\nu_{02}}{1-2\nu_{02}}\cdot\frac{\psi_2(\lambda_2)}{\phi_2(\lambda_2)})r_2^2]\cdot\varepsilon_2 \\ &-\frac{\eta_{02}}{2G_{02}}\frac{\eta_2(\lambda_2)}{\phi_2(\lambda_2)}\alpha_2\tilde{\lambda}_2\end{aligned} \tag{2.95}$$

$$\begin{aligned}\frac{1}{2G_{02}\phi_2(\lambda_2)}\sigma_3^- &= \frac{\nu_{02}}{1-2\nu_{02}}\frac{\psi_2(\lambda_2)}{\phi_2(\lambda_2)}\varepsilon_5 \\ &+\frac{1}{r_3^2-r_2^2}[r_2^2+(1+\frac{2\nu_{02}}{1-2\nu_{02}}\cdot\frac{\psi_2(\lambda_2)}{\phi_2(\lambda_2)})r_3^2]\cdot\varepsilon_3 \\ &-\frac{2r_2^2}{r_3^2-r_2^2}[1+\frac{\nu_{02}}{1-2\nu_{02}}\cdot\frac{\psi_2(\lambda_2)}{\phi_2(\lambda_2)}]\cdot\varepsilon_2 \\ &-\frac{\eta_{02}}{2G_{02}}\frac{\eta_2(\lambda_2)}{\phi_2(\lambda_2)}\alpha_2\tilde{\lambda}_2\end{aligned} \tag{2.96}$$

$$\begin{aligned}\frac{1}{2G_{03}\phi_3(\lambda_3)}\sigma_3^+ &= \frac{\nu_{03}}{1-2\nu_{03}}\frac{\psi_3(\lambda_3)}{\phi_3(\lambda_3)}\varepsilon_5 \\ &+\frac{2r_4^2}{r_4^2-r_3^2}[1+\frac{\nu_{03}}{1-2\nu_{03}}\cdot\frac{\psi_3(\lambda_3)}{\phi_3(\lambda_3)}]\cdot\varepsilon_4 \\ &-\frac{1}{r_4^2-r_3^2}[r_4^2+(1+\frac{2\nu_{03}}{1-2\nu_{03}}\cdot\frac{\psi_3(\lambda_3)}{\phi_3(\lambda_3)})r_3^2]\cdot\varepsilon_3 \\ &-\frac{\eta_{03}}{2G_{03}}\frac{\eta_3(\lambda_3)}{\phi_3(\lambda_3)}\alpha_3\tilde{\lambda}_3\end{aligned} \tag{2.97}$$

$$\frac{1}{2G_{03}\phi_3(\lambda_3)}\sigma_4^- = \frac{\nu_{03}}{1-2\nu_{03}}\frac{\psi_3(\lambda_3)}{\phi_3(\lambda_3)}\varepsilon_5$$
$$+\frac{1}{r_4^2-r_3^2}[r_3^2+(1+\frac{2\nu_{03}}{1-2\nu_{03}}\cdot\frac{\psi_3(\lambda_3)}{\phi_3(\lambda_3)})r_4^2]\cdot\varepsilon_4$$
$$-\frac{2r_3^2}{r_4^2-r_3^2}[1+\frac{\nu_{03}}{1-2\nu_{03}}\cdot\frac{\psi_3(\lambda_3)}{\phi_3(\lambda_3)}]\cdot\varepsilon_3 \tag{2.98}$$
$$-\frac{\eta_{03}}{2G_{03}}\frac{\eta_3(\lambda_3)}{\phi_3(\lambda_3)}\alpha_3\tilde{\lambda}_3$$

Special case: Study of quality and quantity influence of character of physico–chemical change of layers material on stress–strain state of sandwich bodies made up of polymer material is of special interest. Therewith, it is also interesting to determine maximal values of breaking off stresses on the interface between layers depending on change of physico–chemical properties of layer material and arising in the layers of thrust mass forces. In this connection, consider a three-layer pipe made up of a polymer material under the action of influence of corrosive liquid medium. For finding the solution of problem, formula (2.76)–(2.80) are used, in which the external mechanical loads P, p_a, and p_b are equated to zero. We carry out numerical calculations under the following input data. The internal layer of three-layer pipe is made up of polymer material of brand "polyethylene of lower density (PELD), the middle layer", polypropylene (PP), and the external layer polystyrene shock-resistance (PS) with the following geometrical sizes, mechanical properties, and physico–chemical characteristics of layers:

- geometrical sizes of construction of three-layer pipe:

$$r_1 = 0.1\text{m},\ r_2 = 0.13\text{m},\ r_3 = 0.15\text{m},\ r_4 = 0.17\text{m}$$

- mechanical properties and physico–chemical characteristics of internal layer (of polymer PELD):

$$\text{E}_{01} = 34.1667\,\text{MPa},\ \text{E}_{1\min} = 20\text{MPa}\ \ \nu_{01} = 0.4034,\ \nu_{1\max} = 0.4153,$$

$$\lambda_{1\max} = 0.0289,\ \lambda_{1\max}^* = 0.0289,\ \alpha_1 = 0.3452,\ 2\text{G}_{01} = 24.3457\,\text{MPa}$$

$$a_{01} = 50.8336,\ \eta_{01} = 176.8463,\ b_1 = 0.5854,\ k_1 = 0.0119,$$

$$\tilde{\alpha}_1 = \alpha_1 \lambda_{1\max} = 0.01,\ q_1 = 0.0104,\ \phi_1(\lambda_1)\big|_{\tilde{\lambda}_1=1} = 0.5805,$$

$$\psi_1(\lambda_1)\big|_{\tilde{\lambda}_1=1} = 0.6816,\ \eta_1(\lambda_{1\max})\big|_{\tilde{\lambda}_1=1} = 0.0193,\ \gamma_1 = 9025.2\frac{\mathrm{N}}{\mathrm{m}^3}$$

- mechanical properties and physico–chemical characteristics of middle layer (of polymer PP):

$$\mathrm{E}_{02} = 78.3333\,\mathrm{MPa},\ \mathrm{E}_{2\min} = 65\,\mathrm{MPa},\ \nu_{02} = 0.3746,\ \nu_{2\max} = 0.4228,$$

$$\lambda^*_{2\max} = 0.0146,\ \alpha_2 = 0.3,\ 2\mathrm{G}_{02} = 56.9863\,\mathrm{MPa},\ a_{02} = 85.1159,$$

$$\eta_{02} = 312.3337\,\mathrm{MPa},\ b_2 = 0.8298,\ k_2 = 0.0482,\ \tilde{\alpha}_2 = \alpha_2 \lambda_{2\max} = 0.0044,$$

$$q_2 = 0.00524,\ \phi_2(\lambda_2)\big|_{\tilde{\lambda}_2=1} = 0.8017,\ \psi_2(\lambda_{2\max})\big|_{\tilde{\lambda}_2=1} = 1.4698,$$

$$\eta_2(\lambda_{2\max})\big|_{\tilde{\lambda}_2=1} = 0.0197,\ \gamma_1 = 8829\frac{\mathrm{N}}{\mathrm{m}^3}$$

- mechanical properties and physico–chemical characteristics of external layer (of polymer PS):

$$\mathrm{E}_{03} = 153.3333\,\mathrm{MPa},\ \mathrm{E}_{3\min} = 133.8889\,\mathrm{MPa},\ \nu_{03} = 0.33,$$

$$\nu_{3\max} = 0.37,\ \lambda_{3\max} = 0.02,\ \alpha_3 = 0.2333,\ b_3 = 0.8732,\ k_3 = 0.04$$

$$q_3 = 0.0072,\ \tilde{\alpha}_3 = \alpha_2 \lambda_{3\max} = 0.0047,\ 2\mathrm{G}_{03} = 115.2882\,\mathrm{MPa}$$

$$a_{03} = 111.8974,\ \eta_{03} = 450.98,\ \phi_3(\lambda_3)\big|_{\tilde{\lambda}_3=1} = 0.8477,$$

$$\psi_3(\lambda_3)\big|_{\tilde{\lambda}_3=1} = 1.2429,\ \eta_3(\lambda_{3\max})\big|_{\tilde{\lambda}_3=1} = 0.0228,\ \gamma_3 = 10300.5\frac{\mathrm{N}}{\mathrm{m}^3}$$

Therewith, we note the following. By heterogeneity of polymer materials of first, second, and third layers of the pipe, mathematical relation (2.34) in conformity to the considered three-layer pipe, that is, between the swelling parameters of layers λ_1, λ_2, and λ_3, will be in the following form:

$$t^* = \frac{\lambda_1}{q_1} = \frac{\lambda_2}{q_2} = \frac{\lambda_3}{q_3} \tag{2.99}$$

Here q_1, q_2, and q_3 are the coefficients of linear dependences, $0 \le t \le t_{n\max}$ for each layer of a sandwich pipe. They are determined experimentally and a class of polymer materials are represented in Figure 1.3. Formula (2.99) means that for each value of time t changing in the interval ($0 \le t \le t_{n\max}$), the swelling parameters λ_n of the material of each layer will be different. Therefore, while carrying out numerical calculations of specific problems, it is necessary to use the swelling parameters corresponding to the same time. The principle of accordance of these swelling parameter values are represented by relation (2.99). For example, accepting the value $\lambda_3 = \lambda_{3\max}$ for the material of the third layer. Then the corresponding values of parameters λ_1 and λ_2 will be determined from the relation:

$$\lambda_1 = \frac{q_1}{q_3}\lambda_{3\max} = \lambda_1^*, \; \lambda_2 = \frac{q_2}{q_3}\lambda_{3\max} = \lambda_2^* \tag{2.100}$$

Therewith, we note that the values of swelling parameters λ_1^* and λ_2^* are considerably lower than their own maximal values, that is $\lambda_1^* \le \lambda_{1\max}$ and $\lambda_2^* \le \lambda_{2\max}$. In the problem under consideration, the polymer materials have the following maximal values of swelling parameters: $\lambda_{1\max} = 0.035$, $\lambda_{2\max} = 0.02$, $\lambda_{3\max} = 0.02$, and their calculated values equal $\lambda_{1\max}^* = 0.0289$, $\lambda_{2\max}^* = 0.0146$, $\lambda_{3\max}^* = 0.02$. In the considered problem, it is necessary to use the case of minimal values of swelling parameters of layers. The sense of necessity of such calculation is connected with the study of influence of minimal physico–chemical change of layers material on stress distribution in layers. Therefore, the maximal value of the swelling parameter of the third layer is taken as a basis, that is $\lambda^*_{3\max} = 0.02$. The values of swelling parameters of first and second layers were taken from the calculation, that is, $\lambda^*_{1\max} = 0.0289$, $\lambda^*_{2\max} = 0.0146$. If the numerical calculation for maximal values of swelling parameter of all the layers will be carried out, then the calculation will be considerably higher than the obtained ones. Under these experimental data and numerical values of the coefficients, algebraic system (2.76)–(2.80), a_{ij} and b_j for boundary tangential deformation ε_1, ε_2, ε_3, ε_4 and lateral deformation $\varepsilon_5 = \varepsilon_z$, will be represented in the form:

$$\begin{cases} -11 \cdot \varepsilon_1 + 16.9083 \cdot \varepsilon_2 + 0 \cdot \varepsilon_3 + 0 \cdot \varepsilon_4 + \\ 2.4517 \cdot \varepsilon_5 = 0.0834 - 0.0708 \cdot p_a \\ 141.3965 \cdot \varepsilon_1 - 1302.1856 \cdot \varepsilon_2 + 1372.4373 \cdot \varepsilon_3 + \\ 0 \cdot \varepsilon_4 + 90.4533 \cdot \varepsilon_5 = 0.6677 \\ 0 \cdot \varepsilon_1 + 1030.8557 \cdot \varepsilon_2 - 3005.566 \cdot \varepsilon_3 + \\ 2138.6821 \cdot \varepsilon_4 + 13.9773 \cdot \varepsilon_5 = 0.553 \\ 0 \cdot \varepsilon_1 + 0 \cdot \varepsilon_2 - 17.0373 \cdot \varepsilon_3 + 20.8906 \cdot \varepsilon_4 + \\ 1.4231 \cdot \varepsilon_5 = 0.0245 - 0.0102 \cdot p_b \\ -0.693 \cdot \varepsilon_1 - 3.0573 \cdot \varepsilon_2 - 0.623 \cdot \varepsilon_3 + 8.0388 \cdot \varepsilon_4 + \\ 2.8086 \cdot \varepsilon_5 = 0.0338 + 0.318 \cdot P \end{cases} \tag{2.101}$$

For $p_a = p_b = P = 0$, the solution of system (2.101) will be:

$$\varepsilon_1 = -0.0135,\ \varepsilon_2 = -0.0052,\ \varepsilon_3 = -0.0037,$$

$$\varepsilon_4 = -0.0024,\ \varepsilon_5 = 0.0092 \tag{2.102}$$

In this case, numerical values of displacements on the radius of boundary surface points will equal:

$$w\big|_{r_1=0,1\text{m}} = \varepsilon_1 r\big|_{r_1=0,1\text{m}} = -1.35\text{mm},$$

$$w\Big|_{r_2=0.13\text{m}} = \varepsilon_2 r\Big|_{r_2=0.13\text{m}} = -0.676\text{mm},$$

$$w\Big|_{r_3=0.15\text{m}} = \varepsilon_3 r\Big|_{r_3=0.15\text{m}} = -0.555\text{mm},$$

$$w\Big|_{r_4=0.17\text{m}} = \varepsilon_4 r\Big|_{r_4=0.17\text{m}} = -0.408\text{mm} \tag{2.103}$$

Substituting numerical values of tangential and lateral deformations ε_1, ε_2, ε_3, ε_4, $\varepsilon_5 = \varepsilon_z$ (2.102) from formulae (2.87)–(2.92), calculation formulae of stress distribution on the thickness of the first, second, and third layers fill be found. They will be of the form:

For first layer:

$$\begin{pmatrix} \sigma_{yy} \\ \sigma_{rr} \end{pmatrix} = -0.3745 \mp 0.0029 \cdot \frac{1}{r^2} \quad \text{for } 0.1\text{m} \le r_1 \le 0.13\text{m} \tag{2.104}$$
$$\sigma_{zz} = -0.3265$$

for second layer:

$$\begin{pmatrix} \sigma_{yy} \\ \sigma_{rr} \end{pmatrix} = -0.1645 \mp 0.0047 \cdot \frac{1}{r^2} \quad \text{for } 0.13\text{m} \le r_1 \le 0.15\text{m} \tag{2.105}$$
$$\sigma_{zz} = 0.0038$$

for third layer:

$$\begin{pmatrix} \sigma_{yy} \\ \sigma_{rr} \end{pmatrix} = -0.5277 \pm 0.01 \cdot \frac{1}{r^2} \quad \text{for } 0.15\text{m} \le r_1 \le 0.17\text{m} \tag{2.106}$$
$$\sigma_{zz} = 0.215$$

From formula (2.104)–(2.106) define the numerical values of radial stresses at the boundary surface points:

$$\sigma_1^+ = \sigma_r\big|_{r1=0.1\text{m}} = -0.0845\,\text{MPa}\,,$$

$$\sigma_2^- = \sigma_r\big|_{r2=0.13\text{m}} = -0.2029\,\text{MPa}\,,$$

$$\sigma_2^+ = \sigma_r\big|_{r2=0.13\text{m}} = 0.1136\,\text{MPa}\,,$$

$$\sigma_3^- = \sigma_r\big|_{r3=0.15\text{m}} = 0.0444\,\text{MPa}\,,$$

$$\sigma_3^+ = \sigma_r\big|_{r3=0.15\text{m}} = -0.0833\,\text{MPa}\,,$$

$$\sigma_4^- = \sigma_r\big|_{r3=0.15\text{m}} = -0.1817\,\text{MPa}\,,$$

$$(\sigma_{zz})_1 = -0.3265\,\text{MPa}\,,\ (\sigma_{zz})_2 = 0.0038\,\text{MPa}\,,$$

$$(\sigma_{zz})_3 = 0.215\,\text{MPa} \qquad (2.107)$$

From (2.107) it is seen that radial stresses at the boundary surface points of a three-layer situated only under the action of corrosive liquid medium, are in the interval $[(-0.45\frac{\text{kgf}}{\text{cm}^2})–(-1.85\frac{\text{kgf}}{\text{cm}^2})]$. In other words, the arising radial and longitudinal stresses in layers is the result of influence of thrust mass forces arising from physico–chemical change of polymer material of pipe's layers.

Summarizing the above-stated one, we can say that formulae (2.102)–(2.107 completely define the numerical values of displacements and also stresses both on the thickness of pipe's layers and also at the contact points depending on physico–chemical changeability of material of construction's layer.

Therewith, we should note two facts. Firstly, the character of stress strain state of a sandwich pipe, that is, numerical values of displacements, deformations, and stresses in layers will depend on the character of location of polymer layers. Secondly, in the considered problem, the numerical values were obtained for minimal value of swelling parameter of layers. The sense of necessity of such calculation is connected with the study of influence of minimal physico–chemical change of layers material on stress distribution in layers. If the numerical calculation for maximal values of all swelling parameters will be carried out, then these values would be considerably higher than the obtained ones.

2.5 PLANE PROBLEM FOR A PIPE MADE OF POLYMER MATERIAL FASTENED WITH ELASTIC SHELL WITH REGARD TO SWELLING EFFECT

We consider a rather long pipe made up of polymer material rigidly fastened on the endside with a thin-walled linear-elastic shell (Aliyev, 2012). Let this pipe be under the action of internal pressure p_a and force $p_a \cdot a^2$ acting on closed end faces of the shell. Moreover, pressure is transmitted to the sandwich pipe by corrosive liquid. The process of diffusion of corrosive liquid into polymer material occurs under the action of corrosive medium in the internal layer of the sandwich pipe. This, in turn reduces the change of physico–mechanical properties of pipe's internal layer. The main goal of the problem is to investigate influence of swelling effect of polymer material on stress–strain state of the sandwich pipe, and also to define interlayer adhesive strength of the construction (Figure 2.5).

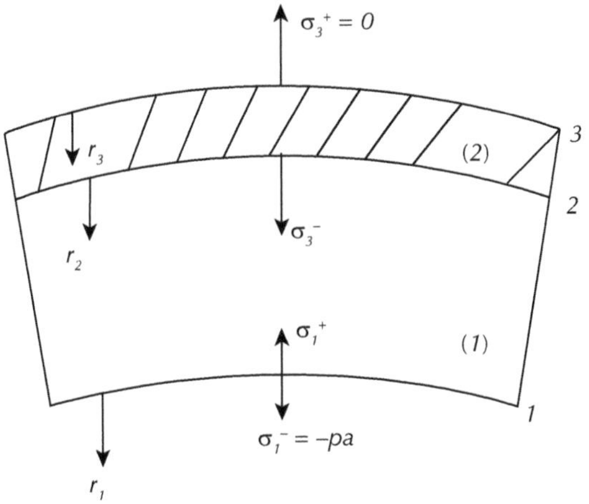

FIGURE 2.5 A pipe made up of polymer material fastened with elastic shell with regard to swelling effect.

Introduce a cylindrical system of coordinates r, y, z and corresponding stress σ_r, σ_y, σ_z and deformation ε_r, ε_y, ε_z components. Moreover,

$$\varepsilon_{rr} = \frac{\partial w}{\partial r},\ \varepsilon_{yy} = \frac{w}{r},\ \varepsilon_z = \frac{\partial u}{\partial z} = \varepsilon_z^0 = const \tag{2.108}$$

Because of a rather long sandwich pipe, that is, the length is considerably greater than the typical size of cross section, there will be a generalized plane deformation on the pipe. Moreover, axial deformation is constant. In this case, the stresses with the index z will be equal to zero, that is, $\sigma_{rz} = \sigma_{yz} = \sigma_{xz} = 0$, $\varepsilon_{rz} = \varepsilon_{yz} = 0$. As it holds diffusion of corrosive liquid medium into polymer material, the cubic deformation of internal pipe under no external loads, $\theta = \varepsilon_{ii} \neq 0$ will be non zero. As the external shell has been made of metal, the corrosive liquid will not diffuse to the external shell.

Under these conditions, Hooke's generalized law with regard to physico–chemical change of polymer material for the internal layer points

$(r_1 \leq r \leq r_2)$ with regard to correction factors of swelling $a_{01}\psi(\lambda)$, $G_{01}\phi(\lambda)$, $\eta_{01}\eta(\lambda)$, and η_{01} (2.109) will be in the form (Aliyev, 2012):

$$\begin{cases} \sigma_{ij} = \grave{a}_{01} \cdot \psi(\lambda) \cdot \theta + 2G_{01} \cdot \phi(\lambda) \cdot \varepsilon_{ij} - \eta_{01} \cdot \eta_1(\lambda) \cdot \alpha \cdot \tilde{\lambda} \cdot \delta_{ij} \\ \sigma = 3a_{01}\psi(\lambda)[1 + \dfrac{2G_{01}}{3a_{01}} \dfrac{\phi(\lambda)}{\psi(\lambda)}](\theta - 3\tilde{\alpha}\lambda) \end{cases} \tag{2.109}$$

In the expanded form $(r_1 \leq r \leq r_2)$:

$$\begin{cases} \sigma_{rr} = \grave{a}_{01} \cdot \psi(\lambda) \cdot \theta + 2G_{01} \cdot \phi(\lambda) \cdot \varepsilon_{rr} - \eta_{01} \cdot \eta(\lambda) \cdot \alpha \cdot \tilde{\lambda} \\ \sigma_{yy} = \grave{a}_{01} \cdot \psi(\lambda) \cdot \theta + 2G_{01} \cdot \phi(\lambda) \cdot \varepsilon_{yy} - \eta_{01} \cdot \eta(\lambda) \cdot \alpha \cdot \tilde{\lambda} \\ \sigma_{zz} = \grave{a}_{01} \cdot \psi(\lambda) \cdot \theta + 2G_{01} \cdot \phi(\lambda) \cdot \varepsilon_{zz} - \eta_{01} \cdot \eta(\lambda) \cdot \alpha \cdot \tilde{\lambda} \end{cases} \tag{2.110}$$

$$\sigma = 3a_{01}\psi(\lambda)[1 + \frac{2G_{01}}{3a_{01}} \frac{\phi(\lambda)}{\psi(\lambda)}](\theta - 3\tilde{\alpha}\lambda) \tag{2.111}$$

Here $\sigma_{ij}, \varepsilon_{ij}$ are stress and strain tensors for the points of first layer $(r_1 \leq r \leq r_2)$, $\theta = \varepsilon_{rr} + \varepsilon_{yy} + \varepsilon_{zz}$ is a cubic deformation, $a_{01} = \dfrac{E_{01}\nu_{01}}{(1+\nu_{01})(1-2\nu_{01})}$ and $2G_{01} = \dfrac{E_{01}}{1+\nu_{01}}$ are the Lame coefficients for the first layer regardless of swelling, that is, $\lambda = 0$;

$$\phi(\lambda) = \frac{b_1^{\tilde{\lambda}}}{1 + \dfrac{k_1}{1+\nu_{01}}\tilde{\lambda}}, \quad \psi(\lambda) = \phi(\lambda)\frac{1 + \dfrac{k_1}{\nu_{01}}\tilde{\lambda}}{1 - \dfrac{2k_1}{1-2\nu_{01}}\tilde{\lambda}},$$

$$\eta(\lambda)=\frac{b_1^{\tilde{\lambda}}\lambda_{\max}}{1-\frac{2k_1}{1-2\nu_{01}}\tilde{\lambda}},\ \eta_{01}=\frac{E_{01}}{1-2\nu_{01}} \tag{2.112}$$

are the correction factors dependent on swelling parameter of layer's internal layer λ; $\tilde{\lambda}=\lambda/\lambda_{\max}$, $k_1=\nu_{1\max}-\nu_{01}$, $b_1=E_{01\max}/E_{01}$; α is a coefficient of linear swelling of internal layer; E_{01}, ν_{01} are elasticity modulus and Poisson ratio for internal layer regardless of swelling, that is, for $\lambda=0$; $\tilde{\alpha}=\alpha\lambda_{\max}$. The change area of correction factor $\varphi(\lambda)$, $\psi(\lambda)$, and $\eta(\lambda)$ will be on the following intervals:

$$1\geq\phi(\tilde{\lambda})\geq\frac{1}{2G_{01}}\cdot\frac{E_{1\min}}{1+\nu_{1\max}},$$

$$1\leq\psi(\tilde{\lambda})\leq\frac{1}{a_{01}}\cdot\frac{E_{1\min}\nu_{1\max}}{(1+\nu_{1\max})(1-2\nu_{1\max})},$$

$$\lambda_{\max}\geq\eta(\tilde{\lambda})\geq\frac{1}{\eta_{01}}\cdot\frac{E_{1\min}\lambda_{1\max}}{1-2\nu_{1\max}} \tag{2.113}$$

The external shell of the sandwich pipe is made of liner-elastic material. Therefore, the physical relation for external layer points of the pipe that change on the interval ($r_2\leq r\leq r_3$) will be of the form:

$$\begin{cases}\sigma_{ij}=a_{02}\cdot\theta\cdot\delta_{ij}+2G_{02}\cdot\varepsilon_{ij} & (2.114)\\ \sigma=3a_{02}(1+\frac{2G_{02}}{3a_{02}})\theta & (2.115)\end{cases}$$

In the expanded form:

$$\begin{cases} \sigma_{rr} = a_{02} \cdot \theta + 2G_{02} \cdot \varepsilon_{rr} \\ \sigma_{yy} = a_{02} \cdot \theta + 2G_{02} \cdot \varepsilon_{yy} \\ \sigma_{zz} = a_{02} \cdot \theta + 2G_{02} \cdot \varepsilon_{zz} \end{cases} \tag{2.116}$$

$$\sigma = 3a_{02}(1 + \frac{2G_{02}}{3a_{02}})\theta \tag{2.117}$$

Here $a_{02} = \frac{E_{02}\nu_{02}}{(1+\nu_{02})(1-2\nu_{02})}$, $2G_{02} = \frac{E_{02}}{1+\nu_{02}}$ are the Lame coefficients for the external linear-elastic layer; E_{02}, G_{02}, ν_{02} are elasticity and shear modulus and Poisson ratio for an external thin walled shell.

Differential equilibrium equations of elementary volume in radial and longitudinal directions will equal:

$$\frac{\partial \sigma_{rr}}{\partial r} = \frac{\sigma_{yy} - \sigma_{rr}}{r} \tag{2.118}$$

$$p_a r_1^2 = \pi \sum_{n=1}^{2} (r_{n+1}^2 - r_n^2)\sigma_{zz}^* \tag{2.119}$$

where, σ_{zz}^* is axial stress in the external thin-walled metallic shell. Substituting (2.110) or (2.116) in equation (2.118), and taking into account (2.108) and boundary conditions of the form:

$$r = r_1,\ w = w_1,$$

$$r = r_2,\ w = w_2, \tag{2.120}$$

represent the deformations of the points of the first layer in thickness by boundary values of radial displacements w_1 and w_2 that will take the form:

$$\begin{pmatrix} \varepsilon_{yy} \\ \varepsilon_{rr} \end{pmatrix} = \frac{(wr)_2 - (wr)_1}{r_2^2 - r_1^2} \pm \frac{1}{r^2} \frac{(w/r)_2 - (w/r)_1}{r_2^{-2} - r_1^{-2}} \tag{2.121}$$

where, w_1 and w_2 are connected with tangential deformations of boundary surfaces ε_1 and ε_2 in the form:

$$\varepsilon_{yy} = \left.\frac{w}{r}\right|_{r=r_1} = \varepsilon_1 \quad \varepsilon_{yy} = \left.\frac{w}{r}\right|_{r=r_2} = \varepsilon_2$$

Substituting (2.121) in (2.110), determine the stresses at the points of the first layer on the thickness $(r_1 \le r \le r_2)$, expressed by boundary tangential deformations ε_1, ε_2 and the longitudinal deformation ε_z in the form:

$$\begin{cases} \dfrac{1}{2G_{01}\phi(\lambda)} \begin{pmatrix} \sigma_{yy} \\ \sigma_{rr} \end{pmatrix} = \dfrac{\nu_{01}}{1-2\nu_{01}} \dfrac{\psi(\lambda)}{\phi(\lambda)} \varepsilon_{zz} \\ +(1 + \dfrac{2\nu_{01}}{1-2\nu_{01}} \dfrac{\psi(\lambda)}{\phi(\lambda)}) \dfrac{\varepsilon_2 r_2^2 - \varepsilon_1 r_1^2}{r_2^2 - r_1^2} \\ \pm \dfrac{1}{r^2} \dfrac{\varepsilon_2 - \varepsilon_1}{r_2^{-2} - r_1^{-2}} - \dfrac{\eta_{01}}{2G_{01}} \dfrac{\eta_1(\lambda)}{\phi(\lambda)} \alpha\tilde{\lambda} & (2.122) \\ \dfrac{1}{2G_{01}\phi(\lambda)} \sigma_{zz} = (1 + \dfrac{\nu_{01}}{1-2\nu_{01}} \dfrac{\psi(\lambda)}{\phi(\lambda)}) \varepsilon_{zz} & (2.123) \\ + \dfrac{2\nu_{01}}{1-2\nu_{01}} \dfrac{\psi(\lambda)}{\phi(\lambda)} \dfrac{\varepsilon_2 r_2^2 - \varepsilon_1 r_1^2}{r_2^2 - r_1^2} \\ - \dfrac{\eta_{01}}{2G_{01}} \dfrac{\eta(\lambda)}{\phi(\lambda)} \alpha\tilde{\lambda} \end{cases}$$

For $r = r_1$ and $r = r_2$, from (2.122) determine the radial stresses on the first and second boundary surfaces, that is, $\sigma_r\big|_{r=r_1} = \sigma_1^+$ and $\sigma_r\big|_{r=r_2} = \sigma_2^-$ which will be equal to:

$$\begin{aligned}\frac{1}{2G_{01}\phi(\lambda)}\sigma_1^+ &= \frac{\nu_{01}}{1-2\nu_{01}}\frac{\psi(\lambda)}{\phi(\lambda)}\varepsilon_{zz}\\ &+\frac{2r_2^2}{r_2^2-r_1^2}\left(1+\frac{\nu_{01}}{1-2\nu_{01}}\frac{\psi(\lambda)}{\phi(\lambda)}\right)\varepsilon_2\\ &-\frac{1}{r_2^2-r_1^2}\left[r_2^2+r_1^2\left(1+\frac{2\nu_{01}}{1-2\nu_{01}}\frac{\psi(\lambda)}{\phi(\lambda)}\right)\right]\varepsilon_1\\ &-\frac{\eta_{01}}{2G_{01}}\frac{\eta(\lambda)}{\phi(\lambda)}\alpha\tilde{\lambda}\end{aligned} \tag{2.124}$$

$$\begin{aligned}\frac{1}{2G_{01}\phi(\lambda)}\sigma_2^- &= \frac{\nu_{01}}{1-2\nu_{01}}\frac{\psi(\lambda)}{\phi(\lambda)}\varepsilon_{zz}\\ &+\frac{1}{r_2^2-r_1^2}\left[r_1^2+r_2^2\left(1+\frac{2\nu_{01}}{1-2\nu_{01}}\frac{\psi(\lambda)}{\phi(\lambda)}\right)\right]\varepsilon_2\\ &-\frac{2r_1^2}{r_2^2-r_1^2}\left(1+\frac{\nu_{01}}{1-2\nu_{01}}\frac{\psi(\lambda)}{\phi(\lambda)}\right)\varepsilon_1-\frac{\eta_{01}}{2G_{01}}\frac{\eta(\lambda)}{\phi(\lambda)}\alpha\tilde{\lambda}\end{aligned} \tag{2.125}$$

Now, substituting (2.121) in (2.117), determines stresses at points of second layer on thickness, expressed by boundary tangential deformations, ε_2, ε_3 and the longitudinal deformation ε_z in the form:

$$\begin{aligned}\frac{1}{2G_{02}}\begin{pmatrix}\sigma_{yy}\\ \sigma_{rr}\end{pmatrix} &= \frac{\nu_{02}}{1-2\nu_{02}}\varepsilon_{zz}+\frac{1}{1-2\nu_{02}}\frac{\varepsilon_3 r_3^2-\varepsilon_2 r_2^2}{r_3^2-r_2^2}\pm\\ &\frac{1}{r^2}\frac{\varepsilon_3-\varepsilon_2}{r_3^{-2}-r_2^{-2}}\end{aligned} \tag{2.126}$$

$$\frac{1}{2G_{02}}\sigma_{zz} = \frac{1-\nu_{02}}{1-2\nu_{02}}\varepsilon_{zz} + \frac{2\nu_{02}}{1-2\nu_{02}}\frac{\varepsilon_3 r_3^2 - \varepsilon_2 r_2^2}{r_3^2 - r_2^2} \tag{2.127}$$

From (2.126) for $r = r_2$ and $r = r_3$ radial stresses on the second and third boundary surfaces, that is, $\sigma_r\big|_{r=r_2} = \sigma_2^+$ and $\sigma_r\big|_{r=r_3} = \sigma_3^-$ will be equal to:

$$\begin{aligned}\frac{1}{2G_{02}}\sigma_2^+ &= \frac{\nu_{02}}{1-2\nu_{02}}\varepsilon_{zz} + \frac{2r_3^2(1-\nu_{02})}{(1-2\nu_{02})(r_3^2 - r_2^2)}\varepsilon_3 \\ &- \frac{r_2^2 + r_3^2(1-2\nu_{02})}{(1-2\nu_{02})(r_3^2 - r_2^2)}\varepsilon_2\end{aligned} \tag{2.128}$$

$$\begin{aligned}\frac{1}{2G_{02}}\sigma_3^- &= -\frac{\nu_{02}}{1-2\nu_{02}}\varepsilon_{zz} + \frac{r_3^2 + r_2^2(1-2\nu_{02})}{(1-2\nu_{02})(r_3^2 - r_2^2)}\varepsilon_3 \\ &- \frac{2r_2^2(1-\nu_{02})}{(1-2\nu_{02})(r_3^2 - r_2^2)}\varepsilon_2\end{aligned} \tag{2.129}$$

Adhesive strength between internal polymer pipe and external linear-elastic metallic shell is determined by equality of radial stresses on the second boundary surface of the form:

$$\sigma_2^+ = \sigma_2^- \tag{2.130}$$

Substitute (2.125) and (2.129) in (2.130) and get:

$$-\frac{G_{01}}{G_{02}}\frac{2r_1^2}{r_2^2 - r_1^2}[\phi(\lambda) + \psi(\lambda)\frac{\nu_{01}}{1-2\nu_{01}}]\varepsilon_1$$

$$+\{\frac{r_2^2+r_3^2(1-2v_{02})}{(r_3^2-r_2^2)(1-2v_{02})}+\frac{G_{01}}{G_{02}}\frac{1}{r_2^2-r_1^2}[(r_1^2+r_2^2)\cdot\phi(\lambda)$$
$$+\psi(\lambda)\frac{2v_{01}}{1-2v_{01}})]\}\varepsilon_2$$

$$-\frac{2r_3^2(1-v_{02})}{(r_3^2-r_2^2)(1-2v_{02})}\varepsilon_3+[\frac{G_{01}}{G_{02}}\frac{v_{01}}{1-2v_{01}}\psi(\lambda)$$
$$-\frac{v_{02}}{1-2v_{02}}]\varepsilon_z=\frac{\eta_{01}}{2G_{02}}\eta(\lambda)\cdot\alpha\cdot\tilde{\lambda} \tag{2.131}$$

The boundary conditions of the problem will be:

$$r=r_1,\ \sigma_1^-=-p_a \tag{2.132}$$

$$r=r_3,\ \sigma_3^+=-p_b\,. \tag{2.133}$$

Contact conditions on the first and third surfaces will be:

$$\sigma_1^+=\sigma_1^-=-p_a \tag{2.134}$$

$$\sigma_3^+=\sigma_3^-=0 \tag{2.135}$$

Substituting (2.124) in (2.134), and (2.129) in (2.135), we get:

$$-\frac{1}{r_2^2-r_1^2}[r_2^2-r_1^2(1-\frac{\psi(\lambda)}{\phi(\lambda)}\frac{2v_{01}}{1-2v_{01}})]\cdot\varepsilon_1$$

$$+\frac{2r_2^2}{r_2^2-r_1^2}\left(1+\frac{\psi(\lambda)}{\phi(\lambda)}\frac{\nu_{01}}{1-2\nu_{01}}\right)\cdot\varepsilon_2 + \frac{\nu_{01}}{1-2\nu_{01}}\frac{\psi(\lambda)}{\phi(\lambda)}\cdot\varepsilon_{zz} = -\frac{p_a-\eta_{01}\eta(\lambda)\alpha\tilde{\lambda}}{2G_{01}\phi(\lambda)} \quad (2.136)$$

$$-\frac{2r_2^2(1-\nu_{02})}{(r_3^2-r_2^2)(1-2\nu_{02})}\varepsilon_2 + \frac{r_3^2+r_2^2(1-2\nu_{02})}{(r_3^2-r_2^2)(1-2\nu_{02})}\varepsilon_3 - \frac{\nu_{02}}{1-2\nu_{02}}\varepsilon_{zz}=0 \quad (2.137)$$

Substituting (2.123) and (2.127) in (2.119), an equation in longitudinal direction of the pipe is defined:

$$-\frac{G_{01}}{G_{02}}\frac{2\nu_{01}}{1-2\nu_{01}}\cdot\psi(\lambda)\cdot r_1^2\cdot\varepsilon_1+\left[\frac{G_{01}}{G_{02}}\frac{2\nu_{01}}{1-2\nu_{01}}\psi(\lambda) - \frac{2\nu_{02}}{1-2\nu_{02}}\right]r_2^2\cdot\varepsilon_2$$

$$+\frac{2\nu_{02}}{1-2\nu_{02}}\cdot r_3^2\cdot\varepsilon_3+\left\{\frac{1-\nu_{02}}{1-2\nu_{02}}(r_3^2-r_2^2) + \left[\frac{G_{01}}{G_{02}}\left(\phi(\lambda)+\psi(\lambda)\frac{\nu_{01}}{1-2\nu_{01}}\right)\right]\cdot(r_2^2-r_1^2)\right\}\cdot\varepsilon_{zz} = \frac{1}{2G_{02}}\left(\frac{r_1^2}{\pi}p_a+\eta_{01}\eta(\lambda)\alpha\tilde{\lambda}\right) \quad (2.138)$$

Thus, we get the following system of four algebraic equations for tangential deformations $\varepsilon_1, \varepsilon_2, \varepsilon_3$ on boundary surfaces and longitudinal deformation $\varepsilon_{zz} = \varepsilon_4$, of the form:

$$-\frac{1}{r_2^2-r_1^2}\left[r_2^2-r_1^2\left(1-\frac{\psi(\lambda)}{\phi(\lambda)}\frac{2\nu_{01}}{1-2\nu_{01}}\right)\right]\cdot\varepsilon_1$$

$$+\frac{2r_2^2}{r_2^2-r_1^2}(1+\frac{\psi(\lambda)}{\phi(\lambda)}\frac{\nu_{01}}{1-2\nu_{01}})\cdot\varepsilon_2$$

$$+\frac{\nu_{01}}{1-2\nu_{01}}\frac{\psi(\lambda)}{\phi(\lambda)}\cdot\varepsilon_{zz}=-\frac{p_a-\eta_{01}\eta(\lambda)\alpha\tilde{\lambda}}{2G_{01}\phi(\lambda)} \tag{2.139}$$

$$-\frac{G_{01}}{G_{02}}\frac{2r_1^2}{r_2^2-r_1^2}[\phi(\lambda)+\psi(\lambda)\frac{\nu_{01}}{1-2\nu_{01}}]\varepsilon_1+$$

$$\{\frac{r_2^2+r_3^2(1-2\nu_{02})}{(r_3^2-r_2^2)(1-2\nu_{02})}+\frac{G_{01}}{G_{02}}\frac{1}{r_2^2-r_1^2}[(r_1^2+r_2^2)\cdot\phi(\lambda)$$

$$+\psi(\lambda)\frac{2\nu_{01}}{1-2\nu_{01}})]\}\varepsilon_2-$$

$$-\frac{2r_3^2(1-\nu_{02})}{(r_3^2-r_2^2)(1-2\nu_{02})}\varepsilon_3+[\frac{G_{01}}{G_{02}}\frac{\nu_{01}}{1-2\nu_{01}}\psi(\lambda)$$

$$-\frac{\nu_{02}}{1-2\nu_{02}}]\varepsilon_{zz}=\frac{\eta_{01}}{2G_{02}}\eta(\lambda)\cdot\alpha\cdot\tilde{\lambda} \tag{2.140}$$

$$-\frac{2r_2^2(1-\nu_{02})}{(r_3^2-r_2^2)(1-2\nu_{02})}\varepsilon_2+\frac{r_3^2+r_2^2(1-2\nu_{02})}{(r_3^2-r_2^2)(1-2\nu_{02})}\varepsilon_3$$

$$-\frac{\nu_{02}}{1-2\nu_{02}}\varepsilon_z=0 \tag{2.141}$$

$$-\frac{G_{01}}{G_{02}}\frac{2\nu_{01}}{1-2\nu_{01}}\cdot\psi(\lambda)\cdot r_1^2\cdot\varepsilon_1+[\frac{G_{01}}{G_{02}}\frac{2\nu_{01}}{1-2\nu_{01}}\psi(\lambda)$$

$$-\frac{2\nu_{02}}{1-2\nu_{02}}]r_2^2\cdot\varepsilon_2+\frac{2\nu_{02}}{1-2\nu_{02}}\cdot r_3^2\cdot\varepsilon_3+$$

$$+\{\frac{1-\nu_{02}}{1-2\nu_{02}}(r_3^2-r_2^2)+[\frac{G_{01}}{G_{02}}(\phi(\lambda)$$

$$+\psi(\lambda)\frac{\nu_{01}}{1-2\nu_{01}})]\cdot(r_2^2-r_1^2)\}\cdot\varepsilon_{zz}$$

$$= \frac{1}{2G_{02}}[\frac{r_1^2}{\pi} p_a + \eta_{01}\eta(\lambda)\alpha\tilde{\lambda}] \tag{2.142}$$

Write system (2.139)–(2.142) in the matrix form:

$$a_{ij} \cdot \varepsilon_j = b_i \tag{2.143}$$

where,

$$\varepsilon_z = \varepsilon_4, \; a_{11} = -\frac{1}{r_2^2 - r_1^2}[r_2^2 - r_1^2(1 - \frac{\psi(\lambda)}{\phi(\lambda)}\frac{2\nu_{01}}{1-2\nu_{01}})]$$

$$a_{12} = \frac{2r_2^2}{r_2^2 - r_1^2}(1 + \frac{\psi(\lambda)}{\phi(\lambda)}\frac{\nu_{01}}{1-2\nu_{01}}), \; a_{13} = 0,$$

$$a_{14} = \frac{\nu_{01}}{1-2\nu_{01}}\frac{\psi(\lambda)}{\phi(\lambda)},$$

$$b_1 = -\frac{p_a - \eta_{01}\eta(\lambda)\alpha\tilde{\lambda}}{2G_{01}\phi(\lambda)},$$

$$a_{21} = -\frac{G_{01}}{G_{02}}\frac{2r_1^2}{r_2^2 - r_1^2}[\phi(\lambda) + \psi(\lambda)\frac{\nu_{01}}{1-2\nu_{01}}],$$

$$a_{22} = \frac{r_2^2 + r_3^2(1-2\nu_{02})}{(r_3^2 - r_2^2)(1-2\nu_{02})}$$

$$+\frac{G_{01}}{G_{02}}\frac{1}{r_2^2 - r_1^2}[(r_1^2 + r_2^2)\cdot\phi(\lambda) + \psi(\lambda)\frac{2\nu_{01}}{1-2\nu_{01}})],$$

$$a_{23} = -\frac{2r_3^2(1-\nu_{02})}{(r_3^2 - r_2^2)(1-2\nu_{02})},$$

$$a_{24} = \frac{G_{01}}{G_{02}}\frac{\nu_{01}}{1-2\nu_{01}}\psi(\lambda) - \frac{\nu_{02}}{1-2\nu_{02}}, \quad b_2 = \frac{\eta_{01}}{2G_{02}}\eta(\lambda)\alpha\tilde{\lambda},$$

$$a_{31} = 0, \quad a_{32} = -\frac{2r_2^2(1-\nu_{02})}{(r_3^2 - r_2^2)(1-2\nu_{02})},$$

$$a_{33} = \frac{r_3^2 + r_2^2(1-2\nu_{02})}{(r_3^2 - r_2^2)(1-2\nu_{02})},$$

$$a_{34} = -\frac{\nu_{02}}{1-2\nu_{02}}, \quad b_3 = 0, \quad a_{41} = -\frac{G_{01}}{G_{02}}\frac{2\nu_{01}}{1-2\nu_{01}}\psi(\lambda)\cdot r_1^2,$$

$$a_{42} = r_2^2[\frac{G_{01}}{G_{02}}\frac{2\nu_{01}}{1-2\nu_{01}}\psi(\lambda) - \frac{2\nu_{02}}{1-2\nu_{02}}], \quad a_{43} = \frac{2\nu_{02}}{1-2\nu_{02}}r_3^2,$$

$$a_{44} = \frac{1-\nu_{02}}{1-2\nu_{02}}(r_3^2 - r_2^2) + [\frac{G_{01}}{G_{02}}(\phi(\lambda)$$
$$+\psi(\lambda)\frac{\nu_{01}}{1-2\nu_{01}})](r_2^2 - r_1^2),$$

$$b_4 = \frac{1}{2G_{02}}[\frac{r_1^2}{\pi}p_a + \eta_{01}\eta(\lambda)\alpha\tilde{\lambda}] \tag{2.144}$$

Having given mechanical characteristics and physico–chemical properties of pipe's layers, and also geometrical characteristics of the construction, the deformations $\varepsilon_1, \varepsilon_2, \varepsilon_3$, $\varepsilon_z = \varepsilon_4$ are uniquely determined from the system of equations (2.139)–(2.142).

Example: Consider stress–strain state of a two-layer pipe consisting of an internal layer made up of polymer material of the brand PS (polystyrene) and external layer made up of steel, under the action of only corrosive oily medium. Investigate the influence of only physico–chemical change of polymer material of the brand polystyrene, and stress–strain state of the two-layer pipe, that is, influence of the swelling effect of the material λ. Carry out numerical calculation for the case of maximal value of the swelling parameter, that is, for the case $\tilde{\lambda} = \frac{\lambda}{\lambda_{\max}} = 1$. Carry out numerical calculations for the following input data:

- geometrical sizes of a two-layer pipe:

  $r_1 = 0.1\text{m}$, $r_2 = 0.13\text{m}$, $r_3 = 0.132\text{m}$;

- internal layer was made of polymer material of the brand PS (polystyrene) with the following mechanical characteristics:

  $\text{E}_{01} = E_{\max} = 153.3333\,\text{MPa}$, $\text{E}_1(\lambda_{\max}) = E_{1\min}(\lambda_{\max}) = 133.8889\,\text{MPa}$

  $\nu_{01} = 0.33$, $\nu_1(\lambda_{\max}) = 0.37$, $\lambda_{\max} = 0.02$, $\alpha_1 = 0.2333$,

  $\gamma_1 = 10{,}300\frac{\text{N}}{\text{m}^3}$, $k_1 = 0.04$, $b_1 = 0.8732$, $2\text{G}_{01} = 115.2882\,\text{MPa}$,

  $q_1 = 0.0072$

- the external layer was made up of steel with mechanical properties:

  $\text{E}_{02} = 196{,}200\,\text{MPa}$, $2\text{G}_{02} = 147{,}518.9\,\text{MPa}$, $\nu_{02} = 0.33$,

  $\lambda_2 = 0$, $\phi_2(\lambda) = \psi_2(\lambda) = 1$, $\psi_1(\lambda) = 1.238$, $\eta_2(\lambda) = 0$

Using the input data, the correction factors $\phi_1(\tilde{\lambda})$, $\psi_1(\tilde{\lambda})$, and $\eta_1(\tilde{\lambda})$ (2.112), dependent on swelling parameter of polymer material for $\tilde{\lambda} = \dfrac{\lambda}{\lambda_{max}} = 1$, will be equal to:

$$\phi_1(\tilde{\lambda})\Big|_{\tilde{\lambda}=1} = 0.8477, \quad \psi_1(\tilde{\lambda})\Big|_{\tilde{\lambda}=1} = 1.238, \quad \eta_1(\tilde{\lambda})\Big|_{\tilde{\lambda}=1} = 0.0227,$$

$$\eta_{01} = 450.8823$$

Substituting these data in system (2.139)–(2.142), we get the following algebraic system for the values of boundary tangential deformations $\varepsilon_1 = \dfrac{w}{r}\Big|_{r=r_1}$, $\varepsilon_2 = \dfrac{w}{r}\Big|_{r=r_2}$, $\varepsilon_3 = \dfrac{w}{r}\Big|_{r=r_3}$, and longitudinal deformation $\varepsilon_4 = \varepsilon_{zz}$ in the form:

$$\begin{cases} -5.1015\varepsilon_1 + 11.87\varepsilon_2 + 0\cdot\varepsilon_3 + 1.4231\varepsilon_4 = 0.0245 \\ -0.0048\varepsilon_1 + 114.2823\varepsilon_2 - 116.5\cdot\varepsilon_3 - 0.9696\varepsilon_4 = 0 \\ 0\cdot\varepsilon_1 + 133.2126\varepsilon_2 + 115\cdot\varepsilon_3 - 0.9706\varepsilon_4 = 0 \\ 0\cdot\varepsilon_1 - 0.0328\varepsilon_2 + 0.0338\varepsilon_3 + 0.001\varepsilon_4 = 0 \end{cases} \tag{2.145}$$

The solution of system (2.145) will be the following numerical values of tangential deformations on the first, second, and third boundary surfaces, and also numerical value of the longitudinal deformation in the form:

$$\varepsilon_1 = -0.0048, \; \varepsilon_2 = \varepsilon_3 = \varepsilon_4 = 0 \tag{2.146}$$

While calculating, the numerical values to the fourth sign inclusively after the comma, were remained. On the basis of the obtained numerical values, we can say that deformation ε_1 is the quantity of thrust deformation arising at the expense of swelling of polymer material of the brand PS from the influence of corrosive liquid medium. In this case, we determine displacement on radius of points of the first boundary surface from:

$$\varepsilon_1\Big|_{r=0,1m} = \frac{w}{r}\Big|_{r_1=0,1m} = -0.0048$$

hence the quantity of displacement on radius w_1 on the first boundary surface will equal to:

$$w_1(\lambda_{\max}) = 0.48\text{mm} \tag{2.147}$$

As for radial displacements on the second and third boundary surface, from equality of deformations to zero $\varepsilon_2 = \varepsilon_3 = \varepsilon_4 = 0$, they will equal to zero, that is,

$$W_2\Big|_{r_2=0,13m} = r_2\Big|_{r_2=0,13} \cdot \varepsilon_2 = 0,$$

$$W_3\Big|_{r_3=0,132m} = r_3\Big|_{r_3=0,132} \cdot \varepsilon_3 = 0 \tag{2.148}$$

Substituting (2.146) in (2.122) and (2.123), define stresses at the points of the first layer, that is, in polymer material, expressed by numerical values of boundary deformations for the value $\tilde{\lambda} = 1$, of the form:

$$\begin{pmatrix} \sigma_{yy} \\ \sigma_{rr} \end{pmatrix} = 0.2432 \pm 0.01 \cdot \frac{1}{r^2} \text{ for } (0.1\text{m} \le r \le 0.13\text{m})$$

$$\sigma_{zz} = -0.43 \tag{2.149}$$

Now, by means of (2.149) define numerical values of radial and tangential stresses on the first and second boundary surfaces, that is, $\sigma_r\big|_{r_1=0,1m}=\sigma_1^+$, $\sigma_r\big|_{r_2=0,13m}=\sigma_2^-$, $\sigma_y\big|_{r=r_1}=\sigma_{y1}^+$, $\sigma_y\big|_{r=r_2}=\sigma_{y2}^-$. They will equal to:

$$\sigma_1^+=-0.7568\,\text{MPa},\ \sigma_{y1}^+\big|_{r1=0.1M}=1.2432\,\text{MPa}$$

$$s_2^-=-0.3485\,\text{MPa},\ \sigma_{y2}^-\big|_{r2=0.13m}=0.8349\,\text{MPa},\ \sigma_{zz}^1=-0.43\,\text{MPa}$$

From (2.126) and (2.127), that allow (2.146), determine the stresses $\sigma_r\big|_{r=r_2}=\sigma_2^+$, $\sigma_r\big|_{r=r_3}=\sigma_3^-$ of second and third metallic layer. As the deformations $\varepsilon_2=\varepsilon_3=\varepsilon_4=0$, stresses $\sigma_r\big|_{r=r_2}=\sigma_2^+$, $\sigma_r\big|_{r=r_3}=\sigma_3^-$ on second and third boundary surface =s will also be equal to zero, that is:

$$\sigma_2^+=\sigma_3^-=0,\ \sigma_{y2}^+=\sigma_{y3}^-=0$$

Thus, numerical values of radial and tangential stresses on the first, second and third boundary surfaces, and also the numerical value of the longitudinal stress in two layer pipe will be:

$$\sigma_1^+=-0.7568\,\text{MPa},\ \sigma_{y1}^+\big|_{r1=0.1m}=1.2432\,\text{MPa}$$

$$\sigma_2^-=-0.3485\,\text{MPa},\ \sigma_{y2}^-\big|_{r2=0.13m}=0.8349\,\text{MPa},$$

$$\sigma_{zz}^1=-0.43\,\text{MPa},$$

$$\sigma_2^+\Big|_{r_2=0.3}=0,\ \sigma_{y2}^+\Big|_{r_2=0.3}=0,\ \sigma_3^-\Big|_{r_3=0.302}=0,\ \sigma_{y3}^-\Big|_{r_3=0.302}=0,$$

$$\sigma_3^+\Big|_{r_3=0.302}=0,\ \sigma_{y3}^+\Big|_{r_3=0.302}=0\ (2.150)$$

Note the following: As all the deformations of the metallic layer equals to zero, then the stresses arising on the second boundary surface as viewed from the first polymer layer, should be balanced by reaction forces R_i, arising on the second boundary surface as viewed from metallic layer. So, the radial stress on the second boundary surface as viewed from the first one, should be balanced with reaction force of quantity (3.55kgf), acting as viewed from the second layer, that is, $R_2=-\sigma_2^- F_{01}=3.55\text{kgf}$

2.6 DYNAMIC BEHAVIOR OF A HOLLOW CYLINDER WITH CHANGING PHYSICO–CHEMICAL PROPERTIES, SITUATED UNDER THE ACTION OF INTERNAL AND EXTERNAL PRESSURES

In this section, we shall give the statement and solution of a rather general problem on quasistatic and dynamic behavior of a thick-walled cylinder made up of polymer material with regard to change of physical–chemical properties of pipe. It is assumed that a polymer pipe or cylinder is rigidly fastened on external surface with chemically stable, that is, not changing its physico–mechanical properties, linear elastic shell. Necessity in investigation of a dynamical problem with regard to the effect of physico–chemical change of polymer material is stipulated by designing of solid propellant rocket engines for purposes of space and rocket engineering, and also by designing of metallic pipes with internal polymer layer for working in corrosive liquid media for the objects of oil chemical and chemical industry.

Below, consider a dynamics problem for the represented general (without details) scheme of rocket engine or of a rather long thick-walled pipe (Figure 2.6). Therewith, the engine's body made up of chemically stable material is a cylindrical shell with which a solid propellant charge is fastened (shaded in the figure).

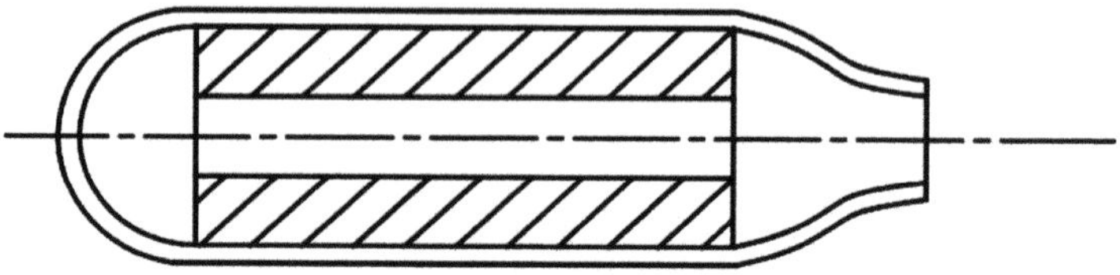

FIGURE 2.6 Scheme of a rocket engine

In the combustion process, the unit state of solid-propellant material will change both at the expense of temperature and pressure factors and physico–chemical properties of fuel material arising because of aggressive medium's influence. In this problem, we'll concentrate our attention on the statement and solution of a dynamics problem with regard to physico–chemical change of solid-propellant medium. One can be acquainted with the character of detailed representation of a solid propellant rocket engine in the papers of J. D. Achenbach (1965), E. S. Sorokin (1967), Ya. M. Shapiro, G. Yu Mazing, N. E. Prudnikov, Shapiro et. al.(1968), A. A. Moskvitin, and V.V. Moskvitin (1972), and also in (Barrer et. al., 1962; Siebel, 1944; Shen et. al., 1976). In this section, we considered a hollow annular cylinder fastened on external surface with a thin-walled cylindrical shell. In such a cylinder, generally speaking, spatial stress stain arises under the action of internal pressure. However, in the cases when the charge's length is considerably greater than the typical size of cross-section, one can use main relations of generalized plane deformation with regard to swelling effect, stated in Chapter 1. Moreover, it is assumed that the quantity of axial deformation is a constant.

Below, we consider a problem in quasi-static and dynamic statement.

2.6.1 *QUASISTATIC BEHAVIOR OF A SANDWICH PIPE*

Introduce a cylindrical system of coordinates r, y, z and corresponding stress and deformation components σ_{rr}, σ_{yy}, σ_{zz}. Moreover:

$$\varepsilon_{rr} = \frac{\partial w}{\partial r},\ \varepsilon_{yy} = \frac{w}{r},\ \text{and}\ \varepsilon_{zz} = \varepsilon = const \tag{2.151}$$

Use the differential balance equation:

$$\frac{d\sigma_{rr}}{dr} = \frac{\sigma_{yy} - \sigma_{rr}}{r} \tag{2.152}$$

and physical relations (2.3) that take into account the influence of change of physico–chemical properties of material (Aliyev, 2012, 2012):

$$\begin{cases} \sigma_{rr} = a_0\psi(\lambda)\cdot\theta + 2G_0\phi(\lambda)\cdot\varepsilon_{rr} - \eta_0\eta(\lambda)\alpha\tilde{\lambda} \\ \sigma_{yy} = a_0\psi(\lambda)\cdot\theta + 2G_0\phi(\lambda)\cdot\varepsilon_{yy} - \eta_0\eta(\lambda)\alpha\tilde{\lambda} \\ \sigma_{zz} = a_0\psi(\lambda)\cdot\theta + 2G_0\phi(\lambda)\cdot\varepsilon_{zz} - \eta_0\eta(\lambda)\alpha\tilde{\lambda} \end{cases} \tag{2.153}$$

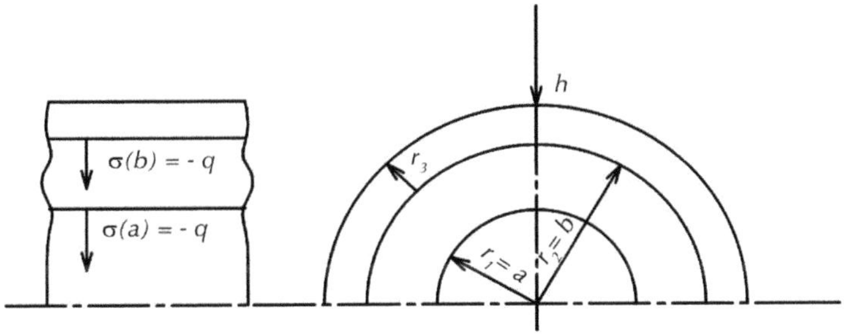

FIGURE 2.7 A thick-walled cylinder made up of polymer material and stiffened with a thin-walled metallic shell

Accept as a basis, the scheme represented in Figure 2.7, write conditions for defining the constants of the equation:

$$\sigma_{rr}(a) = -p,\ \sigma_{rr}(b) = -q$$

$$pa_1^2 = 2\int_a^b \sigma_{zz}(r)rdr + 2\sigma_{zz}^* hb \tag{2.154}$$

where, σ_{rr}^* is an axial stress in the external shell; q is a contact pressure; $h = r_3 - r_2$ is the thickness of external cylinder (Figure2.7). The solution of equation (2.152) will be:

$$\sigma_{rr} = A - \frac{B}{r^2},\ \sigma_{yy} = A + \frac{B}{r^2} \tag{2.155}$$

Find σ_{rr} and ε_{yy}. For that, put together first and second relations of (2.153) and take into account that $\varepsilon_{rr} + \varepsilon_{yy} = \theta - \varepsilon_{zz}$ and from relation (2.15.5) $\sigma_{rr} + \sigma_{yy} = 2A$. In this case, dependence of cubic deformation will be in the form:

$$\theta = \frac{A + \eta_0\eta(\lambda)\alpha\tilde{\lambda} + G_0\phi(\lambda)\varepsilon_z}{a_0\psi(\lambda) + G_0\phi(\lambda)} \tag{2.156}$$

Substituting (2.156) in third relation of (2.153), define the dependence of the longitudinal stress σ_{zz} in the form:

$$\sigma_{zz} = \frac{1}{1 + \frac{G_0}{a_0}\frac{\phi(\lambda)}{\psi(\lambda)}}[A - \frac{G_0}{a_0}\frac{\phi(\lambda)}{\psi(\lambda)}\eta_0\eta(\lambda)\alpha\tilde{\lambda}$$
$$+ G_0\phi(\lambda)(3 + 2\frac{G_0}{a_0}\frac{\phi(\lambda)}{\psi(\lambda)})\cdot\varepsilon_{zz}] \tag{2.157}$$

Now determine the dependence of tangential deformation ε_{yy} on the stresses σ_{rr}, σ_{yy} and σ_{zz}. For that, from (2.153) we have:

$$\varepsilon_{yy} = \frac{1}{2G_0\phi(\lambda)}[\sigma_{yy} + \eta_0\eta(\lambda)\alpha\tilde{\lambda} - a_0\psi(\lambda)\cdot\theta] \tag{2.158}$$

Summing the first and third equations of (2.153) with regard to expression $\varepsilon_{rr}+\varepsilon_{zz}=\theta-\varepsilon_{yy}$, express the cubic deformation θ by stress σ_{rr}, σ_{zz} and the deformation ε_{yy}, in the form:

$$\theta=\frac{1}{2(a_0\psi(\lambda)+G_0\phi(\lambda))}[\sigma_{rr}+\sigma_{zz}+2\eta_0\eta(\lambda)\alpha\tilde{\lambda}$$
$$+2G_0\phi(\lambda)\cdot\varepsilon_{yy}] \tag{2.159}$$

Substituting (2.159) in (2.158) we get:

$$\varepsilon_{yy}=\frac{1}{3G_0\phi(\lambda)}\cdot\frac{1+\frac{G_0}{a_0}\frac{\phi(\lambda)}{\psi(\lambda)}}{1+\frac{2}{3}\frac{G_0}{a_0}\frac{\phi(\lambda)}{\psi(\lambda)}}[\sigma_{yy}$$

$$-\frac{1}{2(1+\frac{G_0}{a_0}\frac{\phi(\lambda)}{\psi(\lambda)})}(\sigma_{rr}+\sigma_{zz})+\frac{\frac{G_0}{a_0}\frac{\phi(\lambda)}{\psi(\lambda)}}{1+\frac{G_0}{a_0}\frac{\phi(\lambda)}{\psi(\lambda)}}\eta_0\eta(\lambda)\alpha\tilde{\lambda}] \tag{2.160}$$

Now, substitute (2.155) and (2.157) in (2.160) and find the dependence of tangential deformation ε_{yy} on the constants A and B, and also on longitudinal deformation ε_{zz} with regard to swelling effect of polymer material λ, in the form:

$$\varepsilon_{yy}=\frac{1}{2G_0\phi(\lambda)(1+\frac{G_0}{a_0}\frac{\phi(\lambda)}{\psi(\lambda)})}[\frac{G_0}{a_0}\frac{\phi(\lambda)}{\psi(\lambda)}\cdot A$$

$$(1+\frac{G_0}{a_0}\frac{\phi(\lambda)}{\psi(\lambda)})\frac{B}{r^2}+\frac{G_0}{a_0}\frac{\phi(\lambda)}{\psi(\lambda)}\eta_0\eta(\lambda)\alpha\tilde{\lambda}-G_0\phi(\lambda)\cdot\varepsilon_{zz}] \tag{2.161}$$

Substituting (2.155) and (2.157) in boundary condition (2.154), we get:

$$\begin{cases} p - q = B\dfrac{M^2 - 1}{r_2^2} \\ p - qM^2 = A(M^2 - 1) \end{cases} \tag{2.162}$$

$$p = \frac{1}{1 + \frac{G_0}{a_0}\frac{\phi(\lambda)}{\psi(\lambda)}}[A - \frac{G_0}{a_0}\frac{\phi(\lambda)}{\psi(\lambda)} \cdot \eta_0 \eta(\lambda)\alpha\tilde{\lambda}$$

$$+G_0\phi(\lambda) \cdot (3 + 2\frac{G_0}{a_0}\frac{\phi(\lambda)}{\psi(\lambda)})\varepsilon_{zz}](M^2 - 1) + 2\sigma_{zz}^* \frac{h}{r_2} M^2 ,$$

$$M = \frac{r_2}{r_1} \tag{2.163}$$

For defining contact pressure q, the condition of conjugation of displacements on the boundary surface between internal polymer pipe and external thin-walled metallic shell is used, in the form:

$$w(r_2) = w_* \tag{2.164}$$

where, w_* is a radial displacement of the shell on the boundary surface $r = r_2$. The following obvious relations for a thin-walled cylindrical shell are known [21]:

$$\begin{cases} \sigma_{yy}^{*} = q\dfrac{r_2}{h}, \sigma_{rr}^{*} = 0 \\ \sigma_{yy}^{*} = \dfrac{1}{1-\nu_*}[\dfrac{E_*}{1+\nu_*}\varepsilon_{yy}^{*} + \dfrac{E_*\nu_*}{1+\nu_*}\varepsilon_{zz}^{*}] \\ \sigma_{zz}^{*} = \nu_*\sigma_{yy}^{*} + E_*\varepsilon_{zz}^{*} \end{cases} \tag{2.165}$$

In this case, by means of condition (2.164) establish the equality of replacements on radius. For that, determine the displacement in an elastic thin-walled shell w_*. From the second equation of (2.165) define tangential deformation on a contact boundary, in the form:

$$\varepsilon_{yy} = \frac{1-\nu_*^2}{E_*}\sigma_{yy} - \nu_*\varepsilon_{zz}^{*} \tag{2.166}$$

Taking into account (2.158) and the first relation of (2.165), the radial displacement at contact point w_* takes the form:

$$w_* = r_2\varepsilon_y^{*} = \frac{qr_2^2}{E_*h}(1-\nu_*^2) - \nu_* r_2\varepsilon_z^{*} \tag{2.167}$$

Here, E_*, ν_* are elasticity modulus and Poisson ratio of the shell's material. From (2.161) and (2.167), equating the expressions of ε_{yy}^{*} for $r = r_2$, we get an additional relation:

$$\frac{A}{2a_0\psi(\lambda)(1+\dfrac{G_0}{a_0}\dfrac{\phi(\lambda)}{\psi(\lambda)})} + \frac{B}{2G_0\phi(\lambda)r_2^2} +$$

$$[\nu_* - \frac{1}{2(1+\dfrac{G_0}{a_0}\dfrac{\phi(\lambda)}{\psi(\lambda)})}]\varepsilon_{zz} = q\frac{r_2}{hE_*}(1-\nu_*^2) -$$

$$\frac{1}{2a_0\psi(\lambda)(1+\frac{G_0}{a_0}\frac{\phi(\lambda)}{\psi(\lambda)})}\cdot\eta_0\eta(\lambda)\alpha\tilde{\lambda} \tag{2.168}$$

Now, from (2.162) and (2.168) determine the constants A and B, and also contact pressure (stress) q, that will be expressed through internal pressure p, mechanical and physico–chemical characteristics of the materials, and the longitudinal deformation ε_z^*, in the form:

$$A=\frac{p-qM^2}{M^2-1},\quad B=\frac{r_2^2}{M^2-1}(p-q) \tag{2.169}$$

$$q=\frac{hE_*}{r_2(1-\nu_*^2)}\{\frac{1}{2a_0\psi(\lambda)(1+\frac{G_0}{a_0}\frac{\phi(\lambda)}{\psi(\lambda)})}\frac{p-qM^2}{M^2-1}$$

$$+\frac{1}{2G_0\phi(\lambda)}\frac{p-q}{M^2-1}+\frac{\eta_0\eta(\lambda)\alpha\tilde{\lambda}}{2a_0\psi(\lambda)(1+\frac{G_0}{a_0}\frac{\phi(\lambda)}{\psi(\lambda)})}$$

$$+[\nu_*-\frac{1}{2(1+\frac{G_0}{a_0}\frac{\phi(\lambda)}{\psi(\lambda)})}]\varepsilon_{zz}^*\} \tag{2.170}$$

Establish the second condition of connections between q and ε_{zz}^*. For that from the first and third relation of (2.165), we have:

$$\sigma_{zz}^*=q\nu_*\cdot\frac{r_2}{h}+E_*\varepsilon_{zz}^* \tag{2.171}$$

Substituting (2.171) in (2.163), define an additional relation between q and ε_{zz}^* in the form:

$$\frac{G_0}{a_0}\frac{\phi(\lambda)}{\psi(\lambda)}p=[G_0\phi(\lambda)\cdot(3+2\frac{G_0}{a_0}\frac{\phi(\lambda)}{\psi(\lambda)})\cdot(M^2-1)+$$
$$2\frac{h}{r_2}E_*(1+\frac{G_0}{a_0}\frac{\phi(\lambda)}{\psi(\lambda)})\cdot M^2]\cdot\varepsilon_{zz}^*$$

$$+qM^2[2\nu_*(1+\frac{G_0}{a_0}\frac{\phi(\lambda)}{\psi(\lambda)})-1]$$
$$-\frac{G_0}{a_0}\frac{\phi(\lambda)}{\psi(\lambda)}(M^2-1)\eta_0\eta(\lambda)\alpha\tilde{\lambda} \tag{2.172}$$

Equations (2.170) and (2.172) will be a system of equations for unknown contact pressure q and longitudinal deformation ε_{zz}^*. From systems (2.170) and (2.172) define the contact pressure q and longitudinal deformation ε_{zz}^* in the form:

$$q=\frac{1}{2G\phi(\lambda)}\frac{1}{M^2-1}\cdot\frac{K_1(p,\lambda)}{K_0(\lambda)},$$

$$\varepsilon_{zz}=\frac{1}{2G\phi(\lambda)}\frac{1}{M^2-1}\cdot\frac{K_2(p,\lambda)}{K_0(\lambda)} \tag{2.173}$$

where, the coefficients $K_0(\lambda)$, $K_1(p,\lambda)$, and $K_2(p,\lambda)$ have the form:

$$K_0(\lambda)=[2\nu_*(1+\frac{G_0}{a_0}\frac{\phi(\lambda)}{\psi(\lambda)})-1][\nu_* -\frac{1}{2(1+\frac{G_0}{a_0}\frac{\phi(\lambda)}{\psi(\lambda)})}]\cdot M^2+[G_0\phi(\lambda)\cdot(3 +2\frac{G_0}{a_0}\frac{\phi(\lambda)}{\psi(\lambda)})(M^2-1)+2\frac{h}{r_2}E_*(1+\frac{G_0}{a_0}\frac{\phi(\lambda)}{\psi(\lambda)})\cdot \cdot M^2]\cdot[\frac{r_2(1-\nu_*^2)}{hE_*} +\frac{1+\frac{G_0}{a_0}\frac{\phi(\lambda)}{\psi(\lambda)}\cdot\left(M^2-1\right)}{2G_0\phi(\lambda)(1+\frac{G_0}{a_0}\frac{\phi(\lambda)}{\psi(\lambda)})(M^2-1)}] \quad (2.174)$$

$$K_1(p,\lambda)=[\frac{1+2\frac{G_0}{a_0}\frac{\phi(\lambda)}{\psi(\lambda)}}{1+\frac{G_0}{a_0}\frac{\phi(\lambda)}{\psi(\lambda)}}\cdot p+\frac{\frac{G_0}{a_0}\frac{\phi(\lambda)}{\psi(\lambda)}}{1+\frac{G_0}{a_0}\frac{\phi(\lambda)}{\psi(\lambda)}} \cdot\eta_0\eta(\lambda)\alpha\tilde{\lambda}]\cdot[G_0\phi(\lambda)\cdot(3+2\frac{G_0}{a_0}\frac{\phi(\lambda)}{\psi(\lambda)})(M^2-1) +2\frac{h}{r_2}E_*(1+\frac{G_0}{a_0}\frac{\phi(\lambda)}{\psi(\lambda)})\cdot M^2]+2G_0\phi(\lambda)\cdot\frac{G_0}{a_0}\frac{\phi(\lambda)}{\psi(\lambda)}[p +(M^2-1)\cdot\eta_0\eta(\lambda)\alpha\tilde{\lambda}]\cdot[\nu_*-\frac{1}{2(1+\frac{G_0}{a_0}\frac{\phi(\lambda)}{\psi(\lambda)})}] \quad (2.175)$$

$$K_2(p,\lambda)=\frac{1}{1+\frac{G_0}{a_0}\frac{\phi(\lambda)}{\psi(\lambda)})}\cdot\{-[(1+2\frac{G_0}{a_0}\frac{\phi(\lambda)}{\psi(\lambda)})\cdot p$$
$$+\frac{G_0}{a_0}\frac{\phi(\lambda)}{\psi(\lambda)}\cdot\eta_0\eta(\lambda)\alpha\tilde{\lambda}]\cdot[2v_*(1$$
$$+\frac{G_0}{a_0}\frac{\phi(\lambda)}{\psi(\lambda)})-1]\cdot M^2+\frac{G_0}{a_0}\frac{\phi(\lambda)}{\psi(\lambda)}[p \tag{2.176}$$
$$+(M^2-1)\cdot\eta_0\eta(\lambda)\alpha\tilde{\lambda}]\cdot[1+\frac{G_0}{a_0}\frac{\phi(\lambda)}{\psi(\lambda)})(M^2+1)$$
$$+2G_0\phi(\lambda)(1+\frac{G_0}{a_0}\frac{\phi(\lambda)}{\psi(\lambda)})(M^2-1)\cdot\frac{r_2(1-v_*^2)}{hE_*}]\}$$

After defining A,B,q and ε_{zz}^* all the stress and strain components become the known functions of pressure p, geometrical parameters r_1,r_2,h, mechanical properties of the elements G_0, a_0, E_*, and also swelling parameter λ. This allows to define the tangential deformation ε_{yy} and stresses σ_{rr}, σ_{yy}, σ_{zz} for each point of a polymer pipe on thickness. For that, by formulae (2.169) and (2.173) substitute the values of A,B,q and ε_{zz}^* in (2.161), (2.155), and (2.157), we get:

$$\varepsilon_{yy}=\frac{1}{(M^2-1)}\cdot\frac{1}{2G_0\phi(\lambda)(1+\frac{G_0}{a_0}\frac{\phi(\lambda)}{\psi(\lambda)})}\{[\frac{G_0}{a_0}\frac{\phi(\lambda)}{\psi(\lambda)}$$
$$+\frac{r_2^2}{r^2}(1+\frac{G_0}{a_0}\frac{\phi(\lambda)}{\psi(\lambda)})]\cdot p+\frac{G_0}{a_0}\frac{\phi(\lambda)}{\psi(\lambda)}(M^2-1)\eta_0\eta(\lambda)\alpha\tilde{\lambda} \tag{2.177}$$
$$-\frac{1}{2}\frac{K_2(p,\lambda)}{K_0(\lambda)}-\frac{1}{2G_0\phi(\lambda)}\frac{1}{(M^2-1)}\cdot[\frac{G_0}{a_0}\frac{\phi(\lambda)}{\psi(\lambda)}M^2$$
$$+(1+\frac{G_0}{a_0}\frac{\phi(\lambda)}{\psi(\lambda)})\frac{r_2^2}{r^2}]\cdot\frac{K_1(p,\lambda)}{K_0(\lambda)}\}$$

$$\sigma_{rr} = \frac{1}{M^2-1}[(1-\frac{r_2^2}{r^2})p - \frac{1}{2G_0\phi(\lambda)}\frac{M^2-\frac{r_2^2}{r^2}}{(M^2-1)}\cdot\frac{K_1(p,\lambda)}{K_0(\lambda)}]$$

$$\sigma_{yy} = \frac{1}{M^2-1}[(1+\frac{r_2^2}{r^2})p - \frac{1}{2G_0\phi(\lambda)}\frac{M^2+\frac{r_2^2}{r^2}}{(M^2-1)}\cdot\frac{K_1(p,\lambda)}{K_0(\lambda)}]$$

$$\begin{aligned}\varepsilon_{yy} = {} & \frac{1}{(M^2-1)}\cdot\frac{1}{2G_0\phi(\lambda)(1+\frac{G_0}{a_0}\frac{\phi(\lambda)}{\psi(\lambda)})}\{[\frac{G_0}{a_0}\frac{\phi(\lambda)}{\psi(\lambda)} \\ & +\frac{r_2^2}{r^2}(1+\frac{G_0}{a_0}\frac{\phi(\lambda)}{\psi(\lambda)})]\cdot p + \frac{G_0}{a_0}\frac{\phi(\lambda)}{\psi(\lambda)}(M^2-1)\eta_0\eta(\lambda)\alpha\tilde{\lambda} \\ & -\frac{1}{2}\frac{K_2(p,\lambda)}{K_0(\lambda)} - \frac{1}{2G_0\phi(\lambda)}\frac{1}{(M^2-1)}\cdot[\frac{G_0}{a_0}\frac{\phi(\lambda)}{\psi(\lambda)}M^2 \\ & +(1+\frac{G_0}{a_0}\frac{\phi(\lambda)}{\psi(\lambda)})\frac{r_2^2}{r^2}]\cdot\frac{K_1(p,\lambda)}{K_0(\lambda)}\}\end{aligned} \quad (2.178)$$

Here, $K_0(\lambda), K_1(p,\lambda)$ and $K_2(p,\lambda)\ q$ have the form of (2.174)–(2.176). Knowing the dependence of change of physico–chemical properties of charge $\varphi(\lambda)$, $\psi(\lambda)$, and $\eta(\lambda)$, formulae (2.177) and (2.178) allow to define tangential deformation at any point on the thickness of charge, and by formula (2.173), the quantity of contact pressure of the charge and thin-walled external shell.

In special case, when the charge's properties donot change the physico–chemical properties, and also neglecting the quantity $\varepsilon_{zz}^* = \varepsilon_{zz}^0$, formulae (2.177) and (2.178) take the form:

$$\varepsilon_{yy}(r)=\frac{p(1+\nu_0)}{E_0(M^2-1)}[1-2\nu_0+\frac{r_2^2}{r^2}$$

$$-\frac{2(1-\nu_0)[(1-2\nu_0)M^2+\frac{r_2^2}{r^2}]}{1+(1-2\nu_0)M^2+\frac{E_0}{E_*}\frac{r_2}{h}(M^2-1)\frac{1-\nu_*^2}{1+\nu_0}}] \tag{2.179}$$

$$\sigma_{rr}(r)=\frac{p}{M^2-1}[1-\frac{r_2^2}{r^2}$$

$$-\frac{2(M^2-\frac{r_2^2}{r^2})(1-\nu_0)}{1+M^2(1-2\nu_0)+\frac{E_0}{E_*}\frac{r_2}{h}(M^2-1)\frac{1-\nu_*^2}{1+\nu_0}}]$$

$$\sigma_{yy}(r)=\frac{p}{M^2-1}[1+\frac{r_2^2}{r^2}$$

$$-\frac{2(M^2+\frac{r_2^2}{r^2})(1-\nu_0)}{1+M^2(1-2\nu_0)+\frac{E_0}{E_*}\frac{r_2}{h}(M^2-1)\frac{1-\nu_*^2}{1+\nu_0}}]$$

$$\sigma_{zz}(r)=\frac{2p\nu_0}{M^2-1}[1-\frac{2M^2(1-\nu_0)}{1+M^2(1-2\nu_0)+\frac{E_0}{E_*}\frac{r_2}{h}(M^2-1)\frac{1-\nu_*^2}{1+\nu_0}}] \tag{2.180}$$

For the case of incompressible solid propellant $\nu_0=\frac{1}{2}$, these formulae are simplified:

$$\varepsilon_{yy}(r)=\frac{p}{E_*}\cdot\frac{r_2^2}{r^2}\cdot\frac{r_2}{h}\cdot\frac{1-\nu_*^2}{1+\frac{2}{3}\frac{E_0}{E_*}\frac{r_2}{h}(M^2-1)(1-\nu_*^2)} \tag{2.181}$$

$$\sigma_{rr}(r) = \frac{p}{M^2-1}[1-\frac{r_2^2}{r^2} - \frac{M^2 - \frac{r_2^2}{r^2}}{1+\frac{2}{3}\frac{E_0}{E_*}\frac{r_2}{h}(M^2-1)(1-\nu_*^2)}]$$

$$\sigma_{yy}(r) = \frac{p}{M^2-1}[1+\frac{r_2^2}{r^2} - \frac{M^2 + \frac{r_2^2}{r^2}}{1+\frac{2}{3}\frac{E_0}{E_*}\frac{r_2}{h}(M^2-1)(1-\nu_*^2)}]$$

$$\sigma_{zz}(r) = \frac{p}{M^2-1}[1- \frac{M^2}{1+\frac{2}{3}\frac{E_0}{E_*}\frac{r_2}{h}(M^2-1)1-\nu_*^2}] \tag{2.182}$$

If the shell is absolutely rigid $(E_0/E_* \to 0)$, then all deformations become equal to zero, the stresses equal: $\sigma_r = \sigma_y = \sigma_z = -p$.

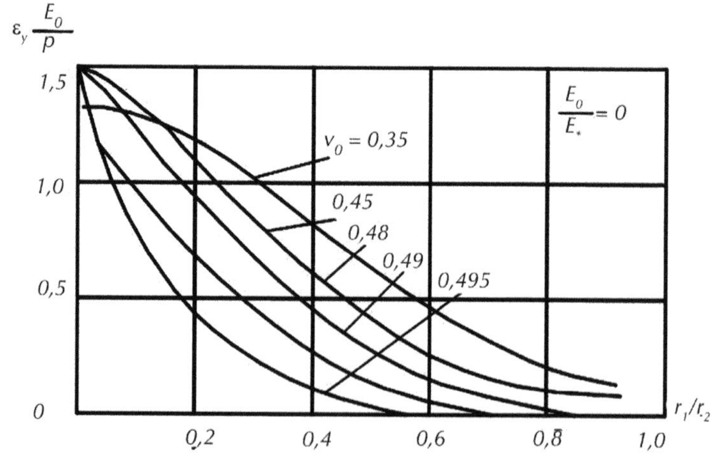

FIGURE 2.8 Annular deformation on internal surface of cylinder

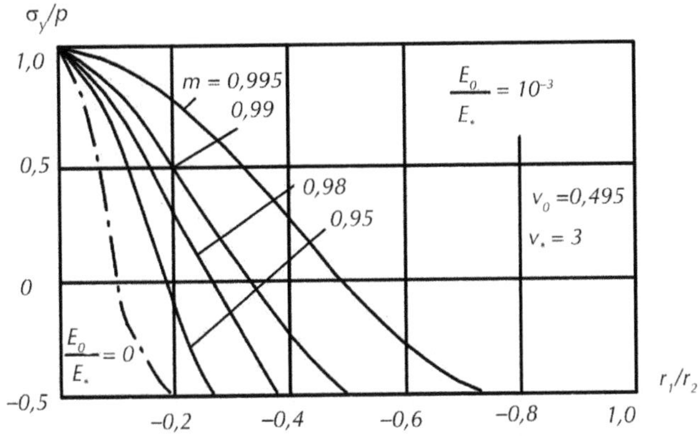

FIGURE 2.9 Plot of annular stress for different values of parameter $m = r_2/(r_2 + h)$

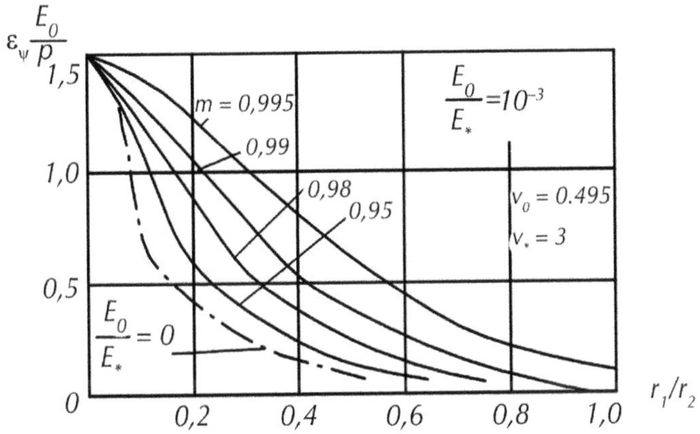

Figure 2.10 Plot of annular deformation for different values of the parameter $m = r_2/(r_2 + h)$

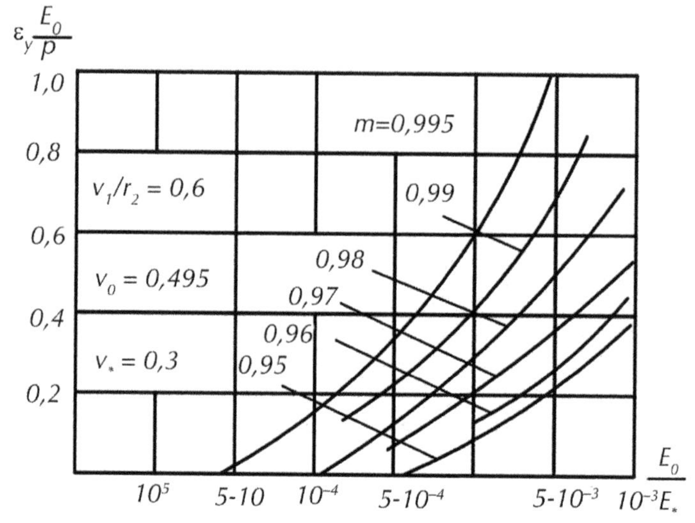

FIGURE 2.11 Influence of parameters E_0/E_* and $m = r_2/(r_2 + h)$ on the value of annular deformation on internal surface of cylinder

The graphs characterizing the influence of separate parameters on stress and deformation of fuel's charge (Summerfield, 1963) are given in figure 2.8–2.11. The graphs are constructed on the basis of formulae (2.179) and (2.180). Graph 2.8 illustrates, how much does the Poisson ratio of the fuel influence on the quantity of annular deformation in the charge's channel for a rigid body. From Figure 2.9, it follows that the annular stress in the channel may be also compressive and stretching whereas the annular deformation is positive for any values of parameters (figure 2.10). The same graphs show the influence of thickness of body h. The dependences of $\varepsilon_y(r_1)$ on the ratio E_0/E_* for different values of λ are illustrated in Figure 2.11. This graph allows to estimate the annular deformation in the channel for different shells (different values of E_* and h).

In the conclusion, note the followings: By carrying out numerical calculation, it was assumed that the properties of fuel E_0 and v_0 are constant and quality and quantity change of stresses and deformations is shown. Taking into account the considerable influence of fuel's properties on problem solution, the quality of change of solution on the degree of change of physico–chemical properties of fuel in the form of swelling λ, and also on temperature, should be expected.

2.6.2 DYNAMICAL BEHAVIOR OF A SANDWICH PLATE

Consider the case of dynamic loading when solid fuel is subjected to linear-elastic law with regard to swelling effect. For that consider solid fuel in the form of an annular cylinder with time variable internal radius $r_1 \leq a(t) \leq r_2$, that is, dynamically loaded with dynamical pressure. The external surface of the cylinder is fastened with elastic thin-walled shell on which external pressure $p_{r_2}(t)$ (Aliyev, 2012, 2012, and Moskvitin, 1972) acts. The solid fuel's material is assumed to be incompressible $\theta = \varepsilon_{ii} = 0$ and the plane deformation conditions are accepted. Therefore, radial displacement $w(r,t)$ and deformations ε_{rr} and ε_{yy} are determined from (2.155) by relations:

$$w(r,t) = \frac{c(r,t)}{r} \tag{2.183}$$

$$\varepsilon_{rr} = -\frac{c(t)}{r^2}, \quad \varepsilon_{yy} = \frac{c(t)}{r^2}, \quad \varepsilon_{zz} = 0 \tag{2.184}$$

The motion equation in radial direction will be:

$$\frac{\partial \sigma_{rr}}{\partial r} + \frac{\sigma_{rr} - \sigma_{yy}}{r} = \rho(\lambda)\frac{\partial^2 w}{\partial t^2} \tag{2.185}$$

where, $\rho = \rho(\lambda)$ is the density of fuel's material with regard to its physico–chemical change. In this case, Hooke's generalized law (2.153) dependent on the parameter of physico–chemical change of material λ in conformity to fuel, will equal (Aliyev, 2012, 2012):

$$\sigma_{rr} = 2G_0\phi(\lambda)\varepsilon_{rr} - \eta_0\eta(\lambda)\alpha\tilde{\lambda}$$

$$\sigma_{yy} = 2G_0\phi(\lambda)\varepsilon_{yy} - \eta_0\eta(\lambda)\alpha\tilde{\lambda}$$

$$\sigma_{zz} = 2G_0\phi(\lambda)\varepsilon_{zz} - \eta_0\eta(\lambda)\alpha\tilde{\lambda} \tag{2.186}$$

Substituting (2.184) in (2.18 6), we find the dependence of stresses σ_{rr} and σ_{yy} on the function $c(t)$, in the form:

$$\sigma_{rr} = -2G_0\phi(\lambda)\frac{c(t)}{r^2} - \eta_0\eta(t)\alpha\tilde{\lambda}$$

$$\sigma_{yy} = 2G_0\phi(\lambda)\frac{c(t)}{r^2} - \eta_0\eta(t)\alpha\tilde{\lambda} \tag{2.187}$$

Hence:

$$\sigma_{yy} - \sigma_{rr} = 4G_0\phi(\lambda)\frac{c(t)}{r^2} \tag{2.188}$$

Substituting (2.188) and (2.183) in (2.185), we get:

$$\frac{\partial\sigma_{rr}}{\partial r} = 4G_0\phi(\lambda)\frac{c(t)}{r^3} + \frac{\rho}{r}\frac{\partial^2 c(t)}{\partial t^2} \tag{2.189}$$

The boundary conditions of the problem will be:

$$\sigma_{rr}(r_1,t) = -p_{r_1}(t) \tag{2.190}$$

$$\sigma_{rr}(r_2,t) = -q(t) = -p_b(t) - \frac{E_* h w(r_2,t)}{r_2^2(1-\nu_*^2)}$$
$$-\rho_* h \frac{\partial^2 w(r_2,t)}{\partial t^2} \quad (2.191)$$

Here q is a contact pressure; ρ_* is material's density; E_*, ν_* are elasticity modulus and Poisson ratio respectively; h is thickness of a linear-elastic thin-walled shell. The second boundary condition (2.191) is written from the known relations for a thin-walled shell, of form:

$$qr_2 = \sigma_y^* h + p_b r_2 + \rho_* \frac{\partial^2 w}{\partial t^2} r_2 h \quad (2.192)$$

$$w_* = \frac{\sigma_y^* r_2}{E_*}(1-\nu_*^2) \quad (2.193)$$

and the condition of continuity of displacements on the contact boundary of form $w(r_2,t) = w_*(t)$.

Further, two problems that will correspond to various boundary values should be distinguished. So, in problem (1), the pressure $p_a(t)$ acts on the internal surface $r_1 = a$, on external surface the pressure equals $p_b(t) = 0$; in problem (2), the internal pressure equals zero, pressure $r_3 = p_b(t)$ acts on the external surface.

The common integral of equation (2.189) will be:

$$\sigma_{rr}(r,t) = -2G_0\phi(\lambda)c(t)\cdot\frac{1}{r^2} + \rho(\lambda)\frac{\partial^2 c(t)}{\partial t^2}\ln r + C \quad (2.194)$$

where, C is an arbitrary constant. Satisfying the first boundary condition (2.190), from relation (2.194) define the constant C, in the form:

$$C = -p_a + 2G_0\phi(\lambda)c(t)\cdot\frac{1}{r_1^2} - \rho(\lambda)\ln r_1 \cdot \frac{\partial^2 c(t)}{\partial t^2} \quad (2.195)$$

Substituting (2.195) in (2.194), the value of radial stress $\sigma_r(r,t)$ for all r changing in the internal $(r_1 \leq r \leq r_2)$ will be of the form:

$$\sigma_r(r,t) = -p_a(t) + 2G_0\phi(\lambda)\frac{r^2 - r_1^1}{r^2 r_1^2}$$
$$+\rho(\lambda)\cdot \ln\frac{r}{r_1}\cdot\frac{\partial^2 c(t)}{\partial t^2} \tag{2.196}$$

Taking into account (2.183), satisfy the boundary condition (2.191), and get the following differential equation:

$$[\rho(\lambda)\ln\frac{b}{a} + \rho_*\frac{h}{r_2}]\cdot\frac{d^2c(t)}{dt^2} + [2G_0\phi(\lambda)\frac{r_2^2 - r_1^2}{r_1^2 r_2^2} +$$
$$\frac{E_* h}{r_2^3(1-\nu_*^2)}]\cdot c(t) = \Delta p(t) \tag{2.197}$$

where:

$$\Delta p(t) = p_a(t) - p_b(t) \tag{2.198}$$

Here ρ_* is the density of external metallic shell, $\rho(\lambda)$ is the density of internal polymer layer dependent on the swelling parameter $_{\Delta p(t)=p_a}$. Note the following two statements of problems:

- Case $\Delta p(t) = p_a(t)$ will correspond to problem (1),
- case $\Delta p(t) = -p_b(t)$ will correspond to problem (2).

The solution of equation (2.197) will be in the form:

$$c(t) = A\sin\left(\sqrt{\frac{2G_0\phi(\lambda)\frac{r_2^2-r_1^2}{r_1^2r_2^2}+\frac{E_*h}{r_2^3(1-\nu_*^2)}}{\rho(\lambda)\ln\frac{b}{a}+\rho_*\frac{h}{r_2}}}t\right)+$$

$$B\cos\left(\sqrt{\frac{2G_0\phi(\lambda)\frac{r_2^2-r_1^2}{r_1^2r_2^2}+\frac{E_*h}{r_2^3(1-\nu_*^2)}}{\rho(\lambda)\ln\frac{b}{a}+\rho_*\frac{h}{r_2}}}t\right)$$

$$+\frac{1}{\sqrt{[\rho(\lambda)\ln\frac{b}{a}+\rho_*\frac{h}{r_2}]\cdot[2G_0\phi(\lambda)\frac{r_2^2-r_1^2}{r_1^2r_2^2}+\frac{E_*h}{r_2^3(1-\nu_*^2)}]}}\int_0^t \Delta p(t-$$

$$\tau)\sin\left(\sqrt{\frac{2G_0\phi(\lambda)\frac{r_2^2-r_1^2}{r_1^2r_2^2}+\frac{E_*h}{r_2^3(1-\nu_*^2)}}{\rho(\lambda)\ln\frac{b}{a}+\rho_*\frac{h}{r_2}}}\tau\right)d\tau \tag{2.199}$$

For finding the constants A and B, we use the initial conditions of form:

$$w(r,0)=0, \quad \left.\frac{\partial w}{\partial t}\right|_{t=0}=0 \tag{2.200}$$

These conditions will correspond to the absence of initial displacements and initial velocities. Taking into account (2.183), the condition on the function $c(t)$ (2.200) will accept the form:

$$\tilde{n}(0)=\left.\frac{dc}{dt}\right|_{t=0}=0 \tag{2.201}$$

Satisfying initial conditions (2.200), find $A = B = 0$. Therefore, the solution (2.198) accepts the form:

$$c(t) = \frac{1}{\sqrt{[\rho(\lambda)\ln\frac{b}{a} + \rho_*\frac{h}{r_2}][2G_0\phi(\lambda)\frac{r_2^2 - r_1^2}{r_1^2 r_2^2} + \frac{E_* h}{r_2^3(1-\nu_*^2)}]}} \cdot$$

$$\int_0^t \Delta p(t-\tau)\sin\left(\sqrt{\frac{2G_0\phi(\lambda)\frac{r_2^2 - r_1^2}{r_1^2 r_2^2} + \frac{E_* h}{r_2^3(1-\nu_*^2)}}{\rho(\lambda)\ln\frac{b}{a} + \rho_*\frac{h}{r_2} - \rho^0}}\tau\right)d\tau \quad (2.202)$$

Let in problem (1), the internal pressure $p_a(t)$ changes in steps from zero to value p_0. In other words, $\Delta p(t) = p_a(t) = p_0 H(t)$, where $H(t)$ is Heaviside function. In this case, from (2.202) we get:

$$c(t) = \frac{p_0}{2G_0\phi(\lambda)\frac{r_2^2 - r_1^2}{r_1^2 r_2^2} + \frac{E_* h}{r_2^3(1-\nu_*^2)}}[1 -$$

$$-\cos\left(\sqrt{\frac{2G_0\phi(\lambda)\frac{r_2^2 - r_1^2}{r_1^2 r_2^2} + \frac{E_* h}{r_2^3(1-\nu_*^2)}}{\rho(\lambda)\ln\frac{b}{a} + \rho_*\frac{h}{r_2}}}t\right)] \quad (2.203)$$

Then, allowing for (2.203) from (2.187) and (2.188), we determine stresses at each point in thickness of internal polymer layer with regard to effect of physico–chemical change of material in form:

$$\sigma_r(r,t) = -p_0 + p_0 \frac{\rho(\lambda)}{\rho(\lambda)\ln\frac{b}{a} + \rho_* \frac{h}{r_2}} \ln\left(\frac{r}{r_1}\right) \cdot$$

$$\cos\left(\sqrt{\frac{2G_0\phi(\lambda)\frac{r_2^2 - r_1^2}{r_1^2 r_2^2} + \frac{E_* h}{r_2^3(1-\nu_*^2)}}{\rho(\lambda)\ln\frac{b}{a} + \rho_* \frac{h}{r_2}}}\, t\right) +$$

$$2G_0\phi(\lambda)\frac{r^2 - r_1^2}{r^2 r_1^2} \frac{\rho(\lambda)\ln\frac{b}{a} + \rho_* \frac{h}{r_2}}{k_0} [1 -$$

$$\cos\left(\sqrt{\frac{2G_0\phi(\lambda)\frac{r_2^2 - r_1^2}{r_1^2 r_2^2} + \frac{E_* h}{r_2^3(1-\nu_*^2)}}{\rho(\lambda)\ln\frac{b}{a} + \rho_* \frac{h}{r_2}}}\, t\right) \tag{2.204}$$

$$\sigma_\phi(r,t) = p_0 + p_0 \frac{\rho(\lambda)}{\rho(\lambda)\ln\frac{b}{a} + \rho_* \frac{h}{r_2}} \ln\left(\frac{r}{r_1}\right) \cdot$$

$$\cos\left(\sqrt{\frac{2G_0\phi(\lambda)\frac{r_2^2 - r_1^2}{r_1^2 r_2^2} + \frac{E_* h}{r_2^3(1-\nu_*^2)}}{\rho(\lambda)\ln\frac{b}{a} + \rho_* \frac{h}{r_2}}}\, t\right) +$$

$$2G_0\phi(\lambda)\frac{r^2 - r_1^2}{r^2 r_1^2} \frac{\rho(\lambda)\ln\frac{b}{a} + \rho_* \frac{h}{r_2}}{2G_0\phi(\lambda)\frac{r_2^2 - r_1^2}{r_1^2 r_2^2} + \frac{E_* h}{r_2^3(1-\nu_*^2)}} [1 -$$

$$\cos\left(\sqrt{\frac{2G_0\phi(\lambda)\frac{r_2^2 - r_1^2}{r_1^2 r_2^2} + \frac{E_* h}{r_2^3(1-\nu_*^2)}}{\rho(\lambda)\ln\frac{b}{a} + \rho_* \frac{h}{r_2}}}\, t\right) \tag{2.205}$$

In the same way, we construct the solution for problem (2) where, $p_a = 0$, and the external pressure changes in steps from zero to the value p^0. In this case,

$$\tilde{n}(t) = -\frac{p^0}{2G_0\phi(\lambda)\frac{r_2^2 - r_1^2}{r_1^2 r_2^2} + \frac{E_* h}{r_2^3(1-\nu_*^2)}}[1 - \cos\left(\sqrt{\frac{2G_0\phi(\lambda)\frac{r_2^2 - r_1^2}{r_1^2 r_2^2} + \frac{E_* h}{r_2^3(1-\nu_*^2)}}{\rho(\lambda)\ln\frac{b}{a} + \rho_*\frac{h}{r_2}}}t\right)$$

$$\sigma_r(t) = -p^0 \frac{\rho(t)}{\rho(\lambda)\ln\frac{b}{a} + \rho_*\frac{h}{r_2}}\ln\left(\frac{r}{r_1}\right)\cdot$$

$$\cos\left(\sqrt{\frac{2G_0\phi(\lambda)\frac{r_2^2 - r_1^2}{r_1^2 r_2^2} + \frac{E_* h}{r_2^3(1-\nu_*^2)}}{\rho(\lambda)\ln\frac{b}{a} + \rho_*\frac{h}{r_2}}}t\right) -$$

$$2G_0\phi(\lambda)\frac{r^2 - r_1^2}{r^2 r_1^2}\frac{p^0}{2G_0\phi(\lambda)\frac{r_2^2 - r_1^2}{r_1^2 r_2^2} + \frac{E_* h}{r_2^3(1-\nu_*^2)}}[1 -$$

$$\cos\left(\sqrt{\frac{2G_0\phi(\lambda)\frac{r_2^2 - r_1^2}{r_1^2 r_2^2} + \frac{E_* h}{r_2^3(1-\nu_*^2)}}{\rho(\lambda)\ln\frac{b}{a} + \rho_*\frac{h}{r_2}}}t\right)]$$

$$\sigma_\phi(t) = -p^0 \frac{\rho(\lambda)}{\rho(\lambda)\ln\frac{b}{a} + \rho_*\frac{h}{r_2}} \ln\left(\frac{r}{r_1}\right) \cdot$$

$$\cos\left(\sqrt{\frac{2G_0\phi(\lambda)\frac{r_2^2 - r_1^2}{r_1^2 r_2^2} + \frac{E_* h}{r_2^3(1-\nu_*^2)}}{\rho(\lambda)\ln\frac{b}{a} + \rho_*\frac{h}{r_2}}}\, t\right) -$$

$$2G_0\phi(\lambda)\frac{r^2 + r_1^2}{r^2 r_1^2} \frac{p^0}{2G_0\phi(\lambda)\frac{r_2^2 - r_1^2}{r_1^2 r_2^2} + \frac{E_* h}{r_2^3(1-\nu_*^2)}} [1 -$$

$$\cos\left(\sqrt{\frac{2G_0\phi(\lambda)\frac{r_2^2 - r_1^2}{r_1^2 r_2^2} + \frac{E_* h}{r_2^3(1-\nu_*^2)}}{\rho(\lambda)\ln\frac{b}{a} + \rho_*\frac{h}{r_2}}}\, t\right) \tag{2.206}$$

In this solution, time variation of contact stress $\sigma_r(r_2,t)$ is of interest. In problem (1), the contact stress is always contractive, in problem (2) it may be also stretching. The numerical solution is represented in Figure 2.12 for $\phi(\lambda) = 1$. In Figure 2.12, the graph of time alternation of contact stress $\sigma_r(r_2,t)$ is given when the external pressure is applied to the charge's body. Therewith, the values of parameters will be the followings: $\rho_*/\rho = 0{,}2$; $h/r_2 = 10^{-2}$; $r_2/r_1 = 3$; and $\phi(\lambda) = 1$ (Akhenbakh, 1965).

It is seen from the graph that in charge in which the elasticity modulus of shell E_* exceeds the shear modulus of fuel G_0 by a factor of 10^4, origination of stretching contact stresses achieving the value $3p^0/4$ that are able to cause peeling of the body from fuel, is possible.

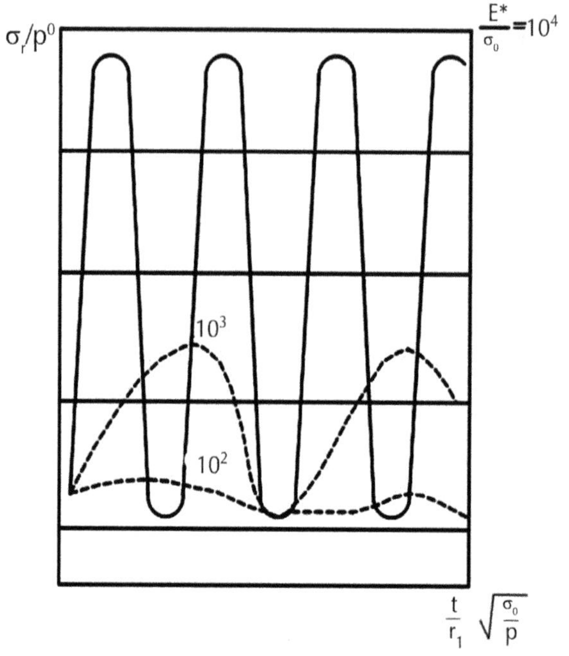

FIGURE 2.12 Time alternation of contact stress for the case of elastic fuel and $\phi(\lambda)=1$

Simultaneously, it should be noted that the influence of swelling function of material $\phi(\lambda)$ may essentially change the peeling of body from the fuel. Therefore, at detailed numerical calculation with regard to the swelling function $\phi(\lambda)$, the maximal value of radial stress may be lower than the value $3p^0/4$.

2.7 KEYWORDS

- **Poisson's ratio**
- **polyethylene of lower density**
- **polypropylene**
- **polystyrene shock-resistance**
- **Sandwich plate**

CHAPTER 3

STRESS, DEFORMATION, AND STRENGTH OF SANDWICH REINFORCED THICK-WALLED PIPES WITH REGARD TO CHANGE OF PHYSICO–CHEMICAL PROPERTIES OF BINDER

G. G. ALIYEV and F. B. NAGIYEV

CONTENTS

3.1 INTERACTION OF REINFORCED ELEMENTS WITH POLYMER MATRIX

A reinforced metal is a composition made of two or more various materials. Application of continuous glass fibers and high-strength polymer fibers for marking goods by the winding method is one of the achievements in the field of choice of reinforced materials. Various thermoactive and thermoplastic resins (e.g., phenolic, polyester, melamine silicon, and epoxy resins) and also high elastic rubber and elastics are used as a binder in reinforced plates. Making of responsible constructions of pipelines, reservoirs, and high-pressure vessels by the winding method is the achievement in the field of application of reinforced plastic materials: Flexibility and high specific strength is the advantage of these goods. Introduction of winding processes to the production of bodies of rocket engines and bodies of buoyant constructions (of tankers, submarines, underwater flexible aerials, and oil storages) accelerated development of manufacturing methods of high strength and reliable goods by the winding method.

In connection with wide application of reinforced plastic materials, today it becomes clear that the basic properties of reinforcement, polymer matrix, and of the whole composition have not been studied enough.

The first successes in production of glass fibers are connected with truly exceptional idea enabling to couple a glass fiber and polymer binder in one material that, separately have fine but opposite structural and technological properties, are the discovery of paramount importance. Such an idea has been borrowed from nature. A great majority of animal and plant fabrics represent themselves as reinforced material.

Today, the reinforced plastic materials with the partially removed main shortcomings of components have been created. However, the question if the attained level of properties of plastic materials is close to ultimate one remains open. The last achievements in creation and study of new fibers say that, the reinforced plastic materials have great potential possibilities. The most effective way of combination of fibers and polymer binders (including finish), and production conditions of compositions are intensively investigated. As a result of these investigations, it now appears that for creating composite materials with optimal properties, it is necessary to take into account the interaction of many factors. Therefore, empirical study of the properties of reinforced plastic materials has given up its place to more thorough experimental, theoretical investigation grounded by experimental test (Aliyev, 1987).

By investigating the composite material, there arise, from mechanics point of view, at least different but closely connected questions. The first question is the estimation of strength of relations between both components on their interface, in other words, provision of adhesive strength of composition that composes a great share of general strength of the reinforced plastic material. The second question is the joint deformation of heterogeneous elements that is a very complicated physico–mechanical process. Physico–chemical change of the material of the elements of a composite under the action of corrosive liquid and gassy media is one of the principle ones in this question. The deformation of the composition, both from purely mechanical position and with regard to influence of physico–chemical change of elements, should be considered first. Mechanical definition of monolith property of heterogeneous system will be first formulated on this basis. The third question is the setting up of quantity relation between the reinforcing substance and polymer binder that provides optimal mechanical properties of the composition in global. Strong coupling of polymer material with reinforcing substance is realized at the expense of interaction forces on their contact surface that are usually called adhesive forces. Physico–chemical nature of these forces have not been completely clarified still, therefore, in references different theories of adhesion can be seen. Some authors suppose that adhesive forces have electrostatic origination, the others suppose chemical bond between the substances to be bound, the third ones takes into account purely mechanical interaction when the binder fills micro non uniformities and pores on the surface of bodies to be bound. But in reality, probably, it holds a complex of mentioned factors. At certain conditions, one or the other of them may take preference. Certainly, the quality relations allowing to estimate adhesive strength by the properties of adhesion, substrate (binder and bound substances) and also influence of change of their physico–chemical properties in course of time, would have the greatest significance for mechanics of reinforced polymers. However, at present, there are no such quality relations in references (Aliyev, 1987, 2012a).

3.2 PROPERTIES OF REINFORCING PENCILS MADE UP OF FIBERS AT QUASISTATIC INTERACTION WITH POLYMER MATRIX

Because of complexity of construction and interaction between the components in heterogeneous media of type of composite and reinforced materials, there is no unique approach for the establishment of their physico–mechanical properties. For instance, in bundle-shaped bodies, the arising forces of interaction with a polymer matrix have a complicated nature that depends both on the kind and degree of heterogeneity of components and on the nature of contact forces, and also on physico–chemical changes of the materials themselves. In this connection, in the existing conceptions and theories of mechanics of composite materials (Aliyev, 1987, 2012), the conjecture of independence of physico–mechanical properties of components on their physico–chemical changes is one of the main acceptable suppositions. *In other words, influence of non mechanical and non temperature character of factors on mechanical properties of polymer materials are not taken into attention.*

In connection with what has been stated, in this chapter we first suggest a way for defining true physico–mechanical properties of reinforced bundles made up of simple fibers of type of filament (braids) and fabrics made of them depending both on the properties of fibers themselves and on arising forces of interaction with polymer matrix of different nature, and also on the effect of change of physico–chemical properties of polymer material.

In the first turn, these forces will depend on physico–chemical properties of a binder, on degree of impregnation and polymerization of the bundle (filament) by a binder, an adhesive bond forces, degree of looseness, twisting and damageability of pencil, on the kind of applied loads and also on change of physico–chemical properties of a polymer material. Mechanical properties of filamentary structures and composite materials depend on such a conglomerate of physical phenomena. In this connection, in this section, experimental–theoretical investigations of author on construction of physico–mechanical models of filamentary structures reinforced in polymer medium and fabrics made of them with regard to influence of these factors should be stated in detail (Aliyev, 1984, 1987, 1995, 1998, 2011, 2012a).

It is known that physico–mechanical properties of simple fibers (e.g., elasticity modulus, Poisson's ratio, ultimate strength, and the curve $\sigma = f(\varepsilon)$) as of any real bodies are independent of conditions of their operation, that is, they will be in the construction's body, for example, the fibers in composite material are same as for free continuous fibers. In other words, they will be defined only by the properties of the bundle made of fibers (e.g., filaments and braids) and fabrics made of them will display essential dependence on many factors, in particular, on stress state of a binder wherein they are reinforced, and generally speaking, they will not coincide with physico–mechanical properties of a bundle while testing them without influence of binding medium. Thus, physico–mechanical properties of a bundle-shaped body will depend both on the properties of the substance of the simple fiber itself and on interaction forces arising between the bundle and binder. Nature of these interaction forces with a polymer matrix will depend firstly on physico–mechanical properties of binder, on degree of impregnation of the bundle by a binder, by adhesive bond forces, degree of looseness and damageability of bundle, on physico–chemical change of polymer material, and also their operation conditions. Therefore, for determining same physico–mechanical properties of bundle shaped bodies with regard to interaction forces between the bundle and medium wherein the bundle is situated, a special definition is introduced (Aliyev, 1984, 1987, 2012a).

Definition: Consider a bundle polymerized into a polymer matrix, and made of fibers (e.g., filament and braid), and consisting of the set of simple fibers densely fastened with a binder. In this case, the longitudinal ε_ℓ^f and lateral $\varepsilon_\perp^f$ deformations of a polymerized bundle will coincide with longitudinal ε_ℓ^b and lateral $\varepsilon_\perp^b$ deformations of a binder. It is supposed that the stress of the bundle situated in binder will depend both on longitudinal deformation of the bundle (filament) ε_ℓ, and on mean values of lateral stresses in the binder $\sigma_\perp^b$ on contact area with the bundle, and on lateral deformation of the reinforcing bundle $\varepsilon_\perp^b$ densely fastened with the binder, in the form:

$$\sigma_\ell = \varphi(\varepsilon_\ell, \varepsilon_\perp^b, \sigma_\perp^b)$$

Representation of specific kinds of deformation modules of a polymerized bundle (e.g., filament and braid) in polymer matrix in the obvious form will depend on the choice of binding material and a bundle made up of fibers. In sections 3.2.1–3.2.4, we shall suggest specific deformation models of polymerized bundles made up of fibers (filaments) with regard to all above mentioned factors, in 3.2.5 we shall suggest deformation modules of a polymerized bundle made up of fibers (filament and braid) with regard to change of physico–chemical properties of the binding material.

3.2.1 ELASTIC PROPERTIES OF A BUNDLE MADE OF FIBERS WITH REGARD TO INTERACTION FORCES WITH ELASTIC MEDIUM

We suggest the following dependence of stress of a polymerized bundle made up of fibers (filament and braid) situated in elastic matrix, on deformation, of the form (Aliyev, 1984, 1987):

$$\sigma_f = T_f / F_f = E_f(\varepsilon_1 + \nu'_\perp \cdot \varepsilon^b_\perp + \frac{\nu''_\perp}{E_f}\sigma^b_\perp) \tag{3.1}$$

where E_f, $\nu'_\perp$, and $\nu''_\perp$ are elastic mechanical properties of a bundle made up of fibers and situated in a polymer matrix, to be experimentally defined; F_f is the cross section area of a polymerized bundle in a polymer matrix; $\varepsilon_\ell = \varepsilon_1$ are the bundle's deformations along the bundle; $\varepsilon^b_\perp$ and $\sigma^b_\perp$ are deformations and stresses lateral to the bundle and are expressed through the main deformations ε^b_2, ε^b_3 and main stresses σ^b_2, σ^b_3, in the form (Figure 3.1):

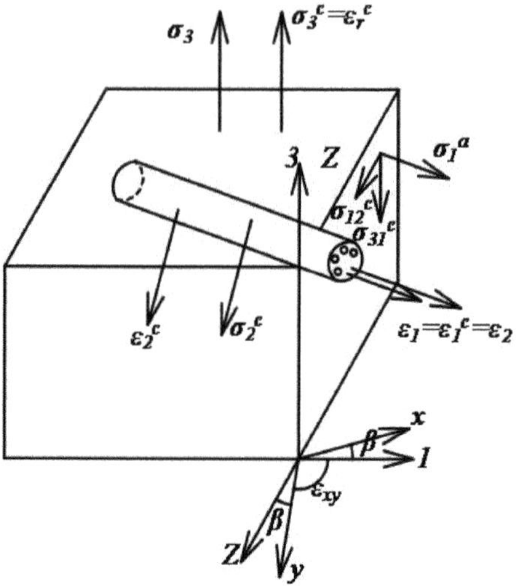

FIGURE 3.1 Interaction of a bundle from fibers with regard to physico–chemical changeability of binder

$$\varepsilon_\perp^b = \frac{1}{2}(\varepsilon_2^b + \varepsilon_3^b), \quad \sigma_\perp^b = \frac{1}{2}(\sigma_2^b + \sigma_3^b) \tag{3.2}$$

Hooke's generalized law for a binder is of the form (Aliyev, 1984, 1987):

$$\begin{cases} E\varepsilon_1^b = \sigma_1^b - \nu_b(\sigma_2^b + \sigma_3^b) = \sigma_1^b - 2\sigma_\perp^b \nu_b \\ E\varepsilon_2^b = \sigma_2^b - \nu_b(\sigma_1^b + \sigma_3^b) \\ E\varepsilon_3^b = \sigma_3^b - \nu_b(\sigma_1^b + \sigma_2^b) \end{cases} \tag{3.3}$$

Where, E_b and ν_b are elasticity modulus and Poisson's ratio respectively, for a binder. Putting together the second and third relations of (3.3), we get:

$$E\varepsilon_\perp^b = (1 - \nu_b)\sigma_\perp^b - \nu_b \sigma_1^b \tag{3.4}$$

Inserting σ_1^b from first relation of (3.3) to (3.4), we get:

$$\sigma_{\perp}^{b} = \frac{E_b}{1 - \nu_b - 2\nu_b^2}(\varepsilon_{\perp}^{b} + \nu_c \varepsilon_1^{b}) \tag{3.5}$$

or

$$\varepsilon_{\perp}^{b} = \frac{1 - \nu_b - 2\nu_b^2}{E_b}\sigma_{\perp}^{c} - \nu_b \varepsilon_1^{b} \tag{3.6}$$

Substituting (3.5) or (3.6) in (3.1), the plot of stress in bundle $\sigma_f = f(\varepsilon_1, \sigma_{\perp}^{b}) = f(\varepsilon_1, \varepsilon_{\perp}^{b})$ situated in a binding medium with regard to arising forces of interaction between the bundle and elastic binder, is expressed in the form:

$$\sigma_f = \frac{T_f}{F_f} = E_f[(1 + \frac{E_b \nu_b}{1 - \nu_b - 2\nu_b^2}\frac{\nu_{\perp}''}{E_f})\varepsilon_1^{b} + (\nu_{\perp}' + \frac{E_b}{1 - \nu_b - 2\nu_b^2}\frac{\nu_{\perp}''}{E_f})\varepsilon_{\perp}^{b}]$$

or

$$\sigma_f = \frac{T_f}{F_f} = E_f[(1 - \nu_b \nu_{\perp}')\varepsilon_1^{b} + \frac{1 - \nu_b - 2\nu_b^2}{E_b}(\nu_{\perp}' + \frac{E_b}{1 - \nu_b - 2\nu_b^2}\frac{\nu_{\perp}''}{E_f})\sigma_{\perp}^{b}] \tag{3.7}$$

Here the elastic constants of polymerized bundle E_f, $\nu_{\perp}'$, and $\nu_{\perp}''$ with regard to the effect of interaction force between bundle and binder should be defined experimentally.

Now represent a model of polymerized bundle in a polymer matrix in a more general form, when the external loads P, p_a, p_b are applied to the body under some angle of ψ to the main axis. Therefore, consider a reinforced shell at whose desired layer, there has been arranged a force layer consisting of twisted bundles (filaments) that slope under the angle ψ to the longitudinal axis (Figure 3.1). Denote the principal axis with respect to the bundle by 1, 2, and 3; normal stresses directed along the normal to the cross-section of the bundle and perpendicular to the bundle by σ_1^b, σ_2^b, and σ_3^b; and the coordinate stresses in the binder by σ_{xx}^b, σ_{yy}^b, and σ_{xy}^b. Bonds of principal stresses in the binder σ_1^b, σ_2^b, and σ_3^b with coordinate stresses for the case of thin-walled shed $\sigma_3^b = 0$, will be expressed by the known relations:

$$\begin{cases} \sigma_1^b = \frac{1}{2}\left(\sigma_{xx}^b + \sigma_{yy}^b\right) + \frac{1}{2}\left(\sigma_{xx}^b - \sigma_{yy}^b\right)\cos 2\psi + \sigma_{xy}^b \sin 2\psi \\ \sigma_2^b = \frac{1}{2}\left(\sigma_{xx}^b + \sigma_{yy}^b\right) - \frac{1}{2}\left(\sigma_{xx}^b - \sigma_{yy}^b\right)\cos\psi - \sigma_{xy}^b \sin 2\psi \\ \sigma_3^b = \frac{1}{2}\left(\sigma_{xx}^b - \sigma_{yy}^b\right)\sin 2\psi + \sigma_{xy}^b \cos 2\psi \end{cases} \tag{3.8}$$

Then on the contact boundary, the lateral stress in the binder $\sigma_\perp^b$ (3.2) allowing (3.8) and $\sigma_3^b = 0$, will be expressed by the coordinate stresses in the form:

$$\sigma_\perp^b = \frac{1}{4}\left(\sigma_{xx}^b + \sigma_{yy}^b\right) - \frac{1}{4}\left(\sigma_{xx}^b - \sigma_{yy}^b\right)\cos 2\psi - \frac{1}{2}\sigma_{xy}^b \sin 2\psi \tag{3.9}$$

On the basis of the third relation of Hooke's generalized law (3.3) and condition $\sigma_3^b = 0$, for the principal deformation ε_3^b of the binder we shall have:

$$\varepsilon_r^b = \varepsilon_3^b = -\frac{\nu}{E}(\sigma_1^b + \sigma_2^b) \tag{3.10}$$

Now, express principal deformations $\varepsilon_1^b, \varepsilon_2^b, \varepsilon_3^b$ by the coordinate deformations $\varepsilon_{xx}^b, \varepsilon_{yy}^b, \varepsilon_{xy}^b$. For that, from the first relation of (3.3) for $\sigma_3^b = 0$, the principal deformation in the binder ε_1^b will be equal to:

$$E\varepsilon_1^b = \sigma_1^b - \nu_b \sigma_2^b \tag{3.11}$$

Substituting the expressions of σ_1^b and σ_2^b from (3.8) in (3.11), we get:

$$\begin{aligned} E_b\varepsilon_1^b &= \frac{1}{2}(1-\nu_b)(\sigma_{xx}^b + \sigma_{yy}^b) \\ &+ \frac{1}{2}(1+\nu_b)(\sigma_{xx}^b - \sigma_{yy}^b)\cdot\cos 2\psi + (1+\nu_b)\sigma_{xy}^b \cdot \sin 2\psi \end{aligned} \tag{3.12}$$

Now use Hooke's law for a shell in the form:

$$\begin{cases} \sigma_{xx}^b = \dfrac{E_b}{1-\nu_b^2}(\varepsilon_{xx}^b + \nu\varepsilon_{yy}^b) \\ \sigma_{yy}^b = \dfrac{E_b}{1-\nu_b^2}(\varepsilon_{yy}^b + \nu\varepsilon_{xx}^b) \\ \sigma_{xy}^b = \dfrac{E_b}{1+\nu_b}\varepsilon_{xy}^b \end{cases} \tag{3.13}$$

From (3.13) we have:

$$\sigma_{xx}^b + \sigma_{yy}^b = \frac{E_b}{1-\nu_b}(\varepsilon_{xx}^b + \varepsilon_{yy}^b),$$

$$\sigma_{xx}^b - \sigma_{yy}^b = \frac{E_b}{1+\nu_b}(\varepsilon_{xx}^b - \varepsilon_{yy}^b) \tag{3.14}$$

Substituting (3.14) in (3.12), the bond of principal deformations $\varepsilon_1^b, \varepsilon_2^b, \varepsilon_3^b$ through the coordinate deformations $\varepsilon_{xx}^b, \varepsilon_{yy}^b, \varepsilon_{xy}^b$ allowing (3.8) for the case a shell, will be represented in the following terminal form:

$$\varepsilon_1^b = \frac{1}{2}\left(\varepsilon_{xx}^b + \varepsilon_{yy}^b\right) + \frac{1}{2}\left(\varepsilon_{xx}^b - \varepsilon_{yy}^b\right)\cos 2\psi + \varepsilon_{xy}^b \sin 2\psi$$

$$\varepsilon_2^b = \frac{1}{2}\left(\varepsilon_{xx}^b + \varepsilon_{yy}^b\right) - \frac{1}{2}\left(\varepsilon_{xx}^b - \varepsilon_{yy}^b\right)\cos 2\psi - \varepsilon_{xy}^b \sin 2\psi$$

$$\varepsilon_3^b = -\frac{\nu_b}{E_b}(\sigma_{xx}^b + \sigma_{yy}^b) \tag{3.15}$$

Substituting (3.15) and (3.14) into the first relation of (3.2) and also (3.14) and (3.13) into (3.9), express the lateral deformation $\varepsilon_\perp^b$ and lateral stress $\sigma_\perp^b$ in the binder through the coordinate deformations in the form:

$$2\varepsilon_\perp^b = \varepsilon_2^b + \varepsilon_3^b = \frac{1-3\nu_b}{2(1-\nu_b)}\left(\varepsilon_{xx}^b + \varepsilon_{yy}^b\right)$$
$$-\frac{1}{2}\left(\varepsilon_{xx}^b - \varepsilon_{yy}^b\right)\cos 2\psi - \varepsilon_{xy}^b \sin 2\psi \tag{3.16}$$

$$2\sigma_\perp^b = \frac{E_b}{2(1-\nu_b)}\left(\varepsilon_{xx}^b + \varepsilon_{yy}^b\right) - \frac{E_b}{2(1+\nu_b)}\left(\varepsilon_{xx}^b - \varepsilon_{yy}^b\right)\cos 2\psi$$
$$-\frac{E_b}{(1+\nu_b)}\varepsilon_{xy}^b \sin 2\psi \tag{3.17}$$

Substituting (3.16) and (3.17) into (3.1), we get a general dependence of elastic stress σ_f of a polymerized bundle made up of fibers on coordinate deformations $\varepsilon_{xx}^{b}, \varepsilon_{yy}^{b}, \varepsilon_{xy}^{b}$ with regard to arising forces of interaction with an elastic matrix, in the form:

$$\sigma_f = \frac{T_f}{F_f}$$

$$= \frac{1}{2} E_f \left[(a + b\cos 2\psi)\varepsilon_{xx}^{b} + (a - b\cos 2\psi)\varepsilon_{yy}^{b} + c\sin 2\psi \cdot \varepsilon_{xy}^{b} \right] \quad (3.18)$$

where,

$$a = 2 + \frac{1-3\nu_b}{1-\nu_b}\nu'_\perp + \frac{E_b}{E_f}\frac{\nu''_\perp}{1-\nu_b}, \quad b = 2 - \nu'_\perp - \frac{E_b}{E_f}\frac{\nu''_\perp}{1+\nu_b},$$

$$c = 2[1 - \nu'_\perp - \frac{E_b}{E_f}\frac{\nu''_\perp}{1+\nu_b}]$$

Special case: In the case of a longitudinally reinforced shell, that is, for $\psi = 0$ wherein $\varepsilon_{ó}^{b}$ is expressed by $\varepsilon_{\perp}^{b}$ by means of (3.16), from (3.18) we get a deformation model of a bundle made up of fibers $\sigma_f = \sigma_f(\varepsilon_1, \varepsilon_{\perp}^{b})$ depending on longitudinal ε_{ℓ} and lateral $\varepsilon_{\perp}^{b}$ deformations with regard to forces of interaction with elastic matrix. But in the case $\psi = 0$ and $\varepsilon_{ó}^{ñ}$ expressed through σ_1^{b}, by means of (3.17), we get a deformation of a polymerized bundle made up of fibers $\sigma_f = \sigma_f(\varepsilon_1, \sigma_{\perp}^{b})$ depending on the longitudinal deformation ε_1 and lateral stress $\sigma_{\perp}^{b}$ with regard to forces of interaction with an elastic matrix, in the form:

$$\sigma_f = T_f / F_f = E_f \left[(1 + \frac{E_b \nu_b}{1 - \nu_b - 2\nu_b^2} \cdot \frac{\nu''_\perp}{E_f})\varepsilon_1 \right.$$

$$\left. + (\nu'_\perp + \frac{E_b}{1 - \nu_b - 2\nu_b^2} \cdot \frac{\nu''_\perp}{E_f})\varepsilon_{\perp}^{b} \right] \quad (3.19)$$

$$\sigma_f = \frac{T_f}{F_f}$$

$$= E_f\left[\left(1-\nu_b\nu'_{\perp}\right)\varepsilon_{\ell} + \frac{1-\nu_b-2\nu_b^2}{E_b}\left(\nu'_{\perp} + \frac{E_b}{1-\nu_b-2\nu_b^2}\frac{\nu''_{\perp}}{E_f}\right)\sigma_{\perp}^b\right] \quad (3.20)$$

These special models coincide with the above suggested models (3.7) and (3.8).

A way for experimental definition of elastic constants of a bundle made up of fibers and situated in the elastic matrix E_f, $\nu'_{\perp}$, and $\nu''_{\perp}$

Elastic constants of boundless made up of fibers (filament and braid) polymerized into the polymer matrix E_f, $\nu'_{\perp}$, and $\nu''_{\perp}$, consisting of filamentary structures should be experimentally defined in the following way: to test two identical compositional samples (one with a bundle, another one without a bundle) in three main directions. Then the property of bundle made up of fibers and situated in polymer medium will be defined by their subtraction, that is

$$P_{\xi} = P'_{\xi} - P''_{\xi}, \quad P_{\eta} = P'_{\eta} - P''_{\eta}, \quad P_{\alpha} = P'_{\alpha} - P''_{\alpha}$$

The compositional samples may be plane either in the form of thick plates or in the form of tubes with longitudinal, annular slanting reinforcement. Below, we shall show one of the variations of experimental definition of elastic constants in the tubes with one-layer force compacted along the sample and consisting of bundles directed along the generator $\psi = 0$. In this case we shall have:

$$\psi = 0, \quad \sigma_1^b = \sigma_{xx}^b, \quad \sigma_2^b = \sigma_{yy}^b, \quad \sigma_3^b = \sigma_{zz}^b = 0,$$

$$\varepsilon_{xy}^b = \sigma_{xy}^b = 0$$

$$\frac{1}{2}(p_a + p_b) = \sigma_3 = \sigma_r, \ \varepsilon_{xx}^b = \varepsilon_\ell = \varepsilon_1, \ \varepsilon_{yy}^b = \varepsilon_2 \tag{3.21}$$

where, p_a and p_b are internal and external pressures; P is the total longitudinal force; m is the amount of bundles made up of fibers per the pipe's cross section; F_f is the area of the bundle made of fibers in the cross section; F_b is the area of the binder in the cross section per filament; F_1 is the cross-section area per polymerized that is, the sum of areas of bundle's cross sections and a filler between the fibers of binder in the form: $F_1 = F_f + F_b$. The total force per section of the reinforced filament will be in form:

$$P = m(T_f + \sigma_1^b F_b) = mE_{averag.} F_1 \varepsilon_1^b \tag{3.22}$$

$$h\sigma_y^b = R(p_a - p_b)$$

$$2\sigma_\perp^b = \sigma_y^b + \sigma_3^b = \frac{R}{h}(p_a - p_b) + \frac{1}{2}(p_a + p_b) \tag{3.23}$$

for thick-walled tubes, and

$$\sigma_\perp^b = \frac{1}{2}\frac{R}{h} p_a, \ p_b = 0 \tag{3.24}$$

for thin-walled tubes. Here, T_f is the force per polymerized bundle from fibers.

Test 1: Simple tension. Let a thin-walled tube with compacted one-directed filaments along the generator be under the action of only tensile force P and $p_a = p_b = 0$. Then:

$$2\sigma_\perp^b = \sigma_y^b, \ \sigma_3^b = 0 \tag{3.25}$$

In this case, the longitudinal deformation ε_1 and applied general longitudinal force P are measured experimentally. The diagram $\frac{P}{mF_1} \sim \varepsilon_1$ is structured on the basis of these measurements. In this case, from the first relation of (3.3) under condition (3.24) and $p_a = 0$, we shall have:

$$\sigma_1^b = E\varepsilon_1^b \tag{3.26}$$

On the other hand, the force per polymerized bundle from fibers T_f will be equal to:

$$T_f = \sigma_f F_f = (EF)_f \cdot \varepsilon_1 \tag{3.27}$$

Substituting (3.26) and (3.27) into (3.22), we get:

$$P/m = \left[(EF)_b + (EF)_f \right] \varepsilon_1 = E_{averag} F_1 \varepsilon_1 \tag{3.28}$$

where, $E_{averag.}$ is a longitudinal elasticity modulus of the composite. The elasticity modulus of the bundle from fibers situated in a n elastic matrix is defined from (3.28), in form:

$$E_f = \frac{1}{k_F}[E_{averag.} - (1 - k_F)E_b],\ k_F = {F_f}/{F_1} \tag{3.29}$$

Test 2: Tension with internal pressure. In second experiment, we use a thin-walled tube with a compacted longitudinally reinforced layer that will be subjected to a test of simultaneous action of internal pressure $p_a = p$, $p_b = 0$ and axial tensile force P. In this case allowing $\sigma_3 = 0$ from (3.23), we have:

$$2\sigma_\perp^b = \left(\frac{R}{h} + \frac{1}{2} \right) p = ap \tag{3.30}$$

The dependence of cubic deformation θ on internal pressure p is structured experimentally:

$$p = K_{averag.}\theta \tag{3.31}$$

where,

$$\theta = \varepsilon_{ii} = \varepsilon_1 + 2\varepsilon_{\perp} \tag{3.32}$$

For finding the constants $\nu'_{\perp}$ and $\nu''_{\perp}$ in the suggested model (3.1), establish bonds between $\varepsilon^b_{\perp}$ and $\sigma^b_{\perp}$. Therewith, we use the experimental data $\frac{P}{mF_1} = f(\varepsilon_1)$ and $p = K_{averag.}\theta$. For that we substitute (3.31) in (3.30), take into account (3.32) and get:

$$\sigma^b_{\perp} = \frac{1}{2} aK_{averag.} \cdot [\varepsilon^b_1 + 2\varepsilon^b_{\perp}] \tag{3.33}$$

or

$$\varepsilon^b_{\perp} = \frac{1}{aK_{averag.}}\sigma^b_{\perp} - \frac{1}{2}\varepsilon^b_1 \tag{3.34}$$

Use the experimental data obtained at longitudinal tension of a tubular sample, that are represented by formula (3.29). The sample's longitudinal tension will be of the form (3.22), that is:

$$P = m(T_f + \sigma^b_1 F_b) = mE_{averag.}F_1\varepsilon_1 \tag{3.35}$$

Substitute σ^b_1 from the Hooke's law (3.3) into (3.35), take into account (3.33) and find the dependence of longitudinal stress in a polymerized bundle from fibers σ_f on longitudinal ε^b_1 and lateral $\varepsilon^b_{\perp}$ deformations, in the form:

$$\sigma_f = {}^{T_f}\!/_{F_f} = \frac{1}{k_F}\{[E_{averag.} - (1-k_F)(E_b + \nu_b aK_{averag.})]\varepsilon_1^b -$$
$$2\nu_b aK_{averag.}(1-k_F)\varepsilon_\perp^b\} \tag{3.36}$$

Here, we used the expression $\sigma_f = k_F \frac{T_f}{F_1}$. In formulae (3.7) and (3.36) compare the coefficients for ε_1^b and $\varepsilon_\perp^b$. Therewith, taking into account formula (3.29), define the dependence of the constants $\nu'_\perp$ and $\nu''_\perp$ on experimental data in the form:

$$\nu'_\perp = \frac{1-2\nu_b}{E_{averag.} - (1-k_F)E_b}(1-k_F)\cdot aK_{averag.}$$

$$\nu''_\perp = -\frac{(1-k_F)}{k_F}\frac{(1+\nu_b)(1-2\nu_b)}{E_b}\cdot aK_{averag.} \tag{3.37}$$

Substituting (3.29) and (3.37) into (3.18), establish the dependence of stress in the bundle from fibers on coordinate deformations with regard to interaction forces with a polymer binder:

$$\sigma_f = \frac{1}{2k_F}\{[(E_{averag.} - (1-k_F)E_b)(1+\cos 2\psi)$$
$$-4\frac{\nu_b(1-2\nu_b)}{1-\nu_b}(1-k_F)\cdot aK_{averag.}]\varepsilon_{xx} +$$

$$[(E_{averag.} - (1-k_F)E_b)(1-\cos 2\psi) +$$
$$2\frac{\nu_b(1-2\nu_b)}{1-\nu_b}(1-k_F)\cdot aK_{averag.}]\varepsilon_{yy} +$$

$$\left[2(E_{averag.}-(1-k_F)E_b)-\frac{1}{2}(1-2\nu_b)(1-k_F)\cdot aK_{averag.}\right]$$
$$\cdot\sin 2\psi\cdot\varepsilon_{xy}\} \quad (3.38)$$

For the case with longitudinal reinforcement $\psi=0$, from (3.38) we shall have the following model:

$$\sigma_f=\frac{1}{k_F}\{[E_{averag.}-(1-k_F)E_b$$
$$-\frac{\nu_b(1-2\nu_b)}{2(1-\nu_b)}(1-k_F)\cdot aK_{averag.}]\varepsilon_{xx}$$

$$-\frac{\nu_b(1-2\nu_b)}{2(1-\nu_b)}(1-k_F)\cdot aK_{averag.}\cdot\varepsilon_{yy}\} \quad (3.39)$$

3.2.2 ELASTICO–PLASTIC PROPERTIES OF A BUNDLE FROM FIBERS WITH REGARD TO INTERACTION FORCE WITH ELASTIC–PLASTIC MATRIX

Let's consider a variant when a bundle from fibers and a matrix are subjected to the theory of small elastic–plastic deformations worked out by A. A. Ilyushin (1948). In this case we suggest the following deformation model of a polymerized bundle from fibers depending on the longitudinal deformation ε_1 and on mean values of lateral stresses $\sigma_\perp^b$ and deformation $\varepsilon_\perp^b$ in binder in form (Aliyev, 1980, 1987):

$$\sigma_f=T_f/F_f=E_f[\varepsilon_1+\nu_\perp'\,\varepsilon_\perp^b+\nu_\perp''\,\frac{\sigma_\perp^b}{E_f}] \quad (3.40)$$

where, E_f, $\nu_\perp'$, and $\nu_\perp''$ are elastico–plastic properties of a polymerized bundle from fibers, dependent both on non linear properties of a polymerized bundle from fibers ω_f, on a binder ω_b and a composite $\omega_{averag.}$. Distinguish in E_f, $\nu'_\perp$, and $\nu''_\perp$ elastic parts and denote them by E_f^e, $\nu'^e_\perp$, $\nu''^e_\perp$, and also introduce the denotation of the following additional functions ω_f, $\Omega'_\perp$, and $\Omega''_\perp$ in the form:

$$E_f = E_f^e(1-\omega_f),\quad \nu'_\perp = \nu'^e_\perp(1-\Omega'_\perp),$$

$$\nu''_\perp = \nu''^e_\perp(1-\Omega''_\perp) \tag{3.41}$$

Then, an elastic–plastic model of a polymerized bundle from fibers in the case of reinforcement will take the following form:

$$\sigma_f = T_f/F_f = E_f^e(1-\omega_f)\cdot[\varepsilon_1 + \nu'^e_\perp(1-\Omega'_\perp)\varepsilon_\perp^b + \nu''^e_\perp(1-\Omega''_\perp)\frac{\sigma_\perp^b}{E_f(1-\omega_f)}] \tag{3.42}$$

Here, E_f^e, $\nu'^e_\perp$, and $\nu''^e_\perp$ are linear-elastic properties of bundle in the matrix, that are defined by formulae (3.29) and (3.37), and the reinforcement functions of A. A. Ilyushin type ω_f, $\Omega'_\perp$, and $\Omega''_\perp$, at first must be expressed through the interaction force between the bundle and polymer matrix; secondly, it is necessary to elaborate a special method for defining them experimentally.

According to A. A. Ilyushin's theory of small elastico–plastic deformations, for a binder in principal stresses and principal deformations, that is, for $i = j$ we shall have (Ilyushin, 1948):

$$\begin{cases} \sigma_{ij} = \sigma\delta_{ij} = \dfrac{2\sigma_i}{3e_i}(\varepsilon_{ij} - \varepsilon\delta_{ij}) & \text{(3.43)} \\ \sigma_i = f(\varepsilon_i) = 3G_b(1-\omega_b)\varepsilon & \text{(3.44)} \\ \sigma = K_b \cdot \theta & \text{(3.45)} \end{cases}$$

for $(i = j, \ldots i, j = 1,2,3)$

In the expanded form:

$$\begin{cases} \sigma_1^b - \sigma^b = \dfrac{2\sigma_i^b}{3\varepsilon_i^b}(\varepsilon_1^b - \varepsilon) \\ \sigma_2^b - \sigma^b = \dfrac{2\sigma_i^b}{3\varepsilon_i^b}(\varepsilon_2^b - \varepsilon) \\ \sigma_3^b - \sigma^b = \dfrac{2\sigma_i^b}{3\varepsilon_i^b}(\varepsilon_3^b - \varepsilon) \end{cases} \tag{3.46}$$

$$\sigma_i = f(\varepsilon_i) = 3G_b(1-\omega_b)\varepsilon \tag{3.47}$$

$$\sigma = K_b \cdot \theta \tag{3.48}$$

where, G_b and K_b are shear modulus and bulk modulus of the material, connected with elasticity modulus E_b and the Poisson ratio ν_b by the expressions:

$$K_b = \frac{E_b}{3(1-2\nu_b)}, \quad G_b = \frac{E_b}{2(1+\nu_b)} \tag{3.49}$$

intensity of stresses σ_i and deformations ε_i are of the form:

$$\sigma_i = \frac{\sqrt{2}}{2}\sqrt{(\sigma_1-\sigma_2)^2+(\sigma_2-\sigma_3)^2+(\sigma_3-\sigma_1)^2} \quad (3.50)$$

$$\varepsilon_i = \frac{\sqrt{2}}{3}\sqrt{(\varepsilon_1-\varepsilon_2)^2+(\varepsilon_2-\varepsilon_3)^2+(\varepsilon_3-\varepsilon_1)^2} \quad (3.51)$$

The mean stress σ^b and cubic deformation θ of the binder will be equal to:

$$\theta = \frac{1}{3}(\varepsilon_1^b+\varepsilon_2^b+\varepsilon_3^b)=\varepsilon\ ,\ \theta = \frac{1}{3}(\varepsilon_1^b+\varepsilon_2^b+\varepsilon_3^b)=\varepsilon \quad (3.52)$$

Introduce the denotation:

$$2\sigma_\perp^b = \sigma_2^b+\sigma_3^b\ ,\ 2\varepsilon_\perp^b = \varepsilon_2^b+\varepsilon_3^b \quad (3.53)$$

From (3.48) having substituted the value $\varepsilon = \frac{1}{3K_b}\sigma$ into the first equation of (3.46) we taking into account (3.52) and (3.53), get:

$$\frac{\sigma_i^b}{\varepsilon_i^b}\varepsilon_1 = (1+\frac{1}{9K_b}\frac{\sigma_i^b}{\varepsilon_i^b})\sigma_1 - (1-\frac{2}{9K_b}\frac{\sigma_i}{\varepsilon_i})\sigma_\perp \quad (3.54)$$

Putting together the second and third equations of (3.46) and taking into account (3.48), (3.52), and (3.53), we get:

$$2\frac{\sigma_i^b}{\varepsilon_i^b}\varepsilon_\perp^b = (1+\frac{4}{9K_b}\frac{\sigma_i^b}{\varepsilon_i^b})\sigma_\perp^b - (1-\frac{2}{9K_b}\frac{\sigma_i^b}{\varepsilon_i^b})\sigma_1^b \quad (3.55)$$

Substituting the value of σ_1 from (3.54) into (3.55), establish a bond between the lateral deformational $\varepsilon_\perp^b$ depending on the lateral stress $\sigma_\perp^b$, and longitudinal deformation ε_1^b in the form:

$$2(9K_b + \frac{\sigma_i^b}{\varepsilon_i^b})\varepsilon_\perp^b = 9\sigma_\perp^b - (9K_b - 2\frac{\sigma_i^b}{\varepsilon_i^b})\varepsilon_1^b \qquad (3.56)$$

Substituting the expression of $\varepsilon_\perp^b$ from (3.56) into (3.42) and also substituting the expression of $\sigma_\perp^b$ from (3.56) into (3.42), we establish the dependence of the stress $\sigma_f = \frac{T_f}{F_f}$ of a bundle from fibers polymerized to elastic–plastic matrix both on ε_1, $\varepsilon_\perp^b$, and ε_1 and on $\sigma_\perp^b$ with regard to arising interaction forces between the bundle from fibers and the binder in the form:

$$\sigma_f = \frac{T_f}{F_f} = E_f^e(1-\omega_f)\{[(1 - \frac{9K_b - 2\frac{\sigma_i^b}{\varepsilon_i^b}}{2(9K_b + \frac{\sigma_i^b}{\varepsilon_i^b})}\nu_\perp']\varepsilon_1^b$$

$$+[\frac{9}{2(9K_b + \frac{\sigma_i^b}{\varepsilon_i^b})}\nu_\perp' + \frac{1}{E_f^e(1-\omega_f)}\nu_\perp'']\sigma_\perp^b\} \qquad (3.57)$$

$$\sigma_f = \frac{T_f}{F_f} = E_f^e(1-\omega_f)\{[(1 + \frac{9K_b - 2\frac{\sigma_i^b}{\varepsilon_i^b}}{9E_f^e(1-\omega_f)}\nu_\perp'']\varepsilon_1^b$$

$$+[\nu_\perp' + \frac{2(9K_b + \frac{\sigma_i^b}{\varepsilon_i^b})}{9E_f^e(1-\omega_f)}\nu_\perp'']\varepsilon_\perp^b\} \qquad (3.58)$$

Consider the case when a binder is subjected to the linearly strengthening law (Ilyushin, 1948; Ilyushin and Lenskiy, 1959):

$$\sigma_i^b = 3G_b(1-\omega_b)\varepsilon_i^b \qquad (3.59)$$

Substituting (3.59) into (3.57) and (3.58), taking into account (3.49) and $E_b = \frac{9K_b G_b}{3K_b + G_b}$ in the terminal form, we get:

$$\sigma_f = E_f^e(1-\omega_f)\{[(1-\frac{3\nu_b + (1-2\nu_b)\omega_b}{3-(1-2\nu_b)\omega_b}\nu_\perp^{'e}(1-\Omega_\perp^{'})]\varepsilon_1^b$$
$$+[\frac{3}{E_b}\frac{(1-2\nu_b)(1+\nu_b)}{3-(1-2\nu_b)\omega_b}\nu_\perp^{'e}(1-\Omega_\perp^{'})$$
$$+\frac{1}{E_f^e(1-\omega_f)}\nu_\perp^{''e}(1-\Omega_\perp^{''})]\sigma_\perp^b\} \tag{3.60}$$

$$\sigma_f = E_f^e(1-\omega_f)\{[(1$$
$$+\frac{E_b}{3E_f^e(1-\omega_f)}\frac{3\nu_b + (1-2\nu_b)\omega_b}{(1-2\nu_b)(1+\nu_b)}\nu_\perp^{''e}(1-\Omega_\perp^{''})]\varepsilon_1^b$$
$$+[\nu_\perp^{'e}(1-\Omega_\perp^{'}) + \frac{E_b}{3E_f^e(1-\omega_f)}\frac{3-(1-2\nu_b)\omega_b}{(1-2\nu_b)(1+\nu_b)}\nu_\perp^{''e}(1-\Omega_\perp^{''})]\varepsilon_\perp^b\} \tag{3.61}$$

Experimental–theoretical method for defining elastico–plastic bundles from fibers situated in elastic–plastic matrix

$$E_f = E_f^e(1-\omega_f),\ \omega_f,\ \nu_\perp' = \nu_\perp'^e(1-\Omega_\perp'),\ \nu_\perp'' = \nu_\perp''^e(1-\Omega_\perp'')$$

For defining elastico–plastic properties of a bundle from fibers with regard to interaction forces with elastic–plastic binder, we suggest the following experimental–theoretical method. Let's consider a tube with compacted along the sample one layer reinforced force layer consisting of bundles from fibers directed along the generator. In this case we shall have:

$$\sigma_1^b = \sigma_x^b,\ \sigma_2^b = \sigma_y^b,\ \sigma_{xy}^b = 0,\ \sigma_3^b = 0,$$

$$\sigma_r^b = \sigma_3^b = \frac{p_a - p_b}{2},\ \sigma_3^b = 0,$$

$$\sigma_{\perp}^{b}=\sigma_{2}^{b}=\frac{R}{h}(p_a-p_b)\quad \varepsilon_1^b=\varepsilon_x,\quad \varepsilon_2^b=\varepsilon_y^b,\quad \varepsilon_{xy}^b=0$$

Test 1: Simple tension. In this case, a composite tube is situated only under the action of tensile force P, and $p_a=p_b=0$. Then:

$$\sigma_3^b=0,\quad \sigma_y^b=\frac{R}{h}(p_a-p_b)=0,\quad \sigma_{\perp}^b=\frac{1}{2}\frac{R}{h}p_a=0,\quad p_b=0 \tag{3.62}$$

In this case, we construct the experimental diagram $P/mF_1=f(\varepsilon_1)=E_{averag.}^b(1-\omega_{averag.})\varepsilon_1$ that has a non linear form:

$$P=m(T_f+\sigma_1F_b')=E_{averag.}(1-\omega_{averag.})F_1m\varepsilon_1 \tag{3.63}$$

Here, $\omega_{averag.}$ is the longitudinal reinforcement coefficient of A. A. Ilyushin's type material. From (3.54) for $\sigma_{\perp}=0$, we have the following relationship between ε_1^b and σ_1^b:

$$\sigma_1^b=\frac{9K_b\dfrac{\sigma_i}{\varepsilon_i}}{9K_b+\dfrac{\sigma_i}{\varepsilon_i}}\cdot\varepsilon_1^b \tag{3.64}$$

Substituting (3.64) into (3.63), we get:

$$E_f^e(1-\omega_f)=\frac{1}{k_F}\left[E_{averag.}(1-\omega_{averag.})-(1-k_F)\frac{9K_b\dfrac{\sigma_i^b}{\varepsilon_i^c}}{9K_b+\dfrac{\sigma_i^b}{\varepsilon_i^b}}\right] \tag{3.65}$$

For distinguishing the linear elastic part in the right hand side and for defining the form of the function ω_f, we set the function $\sigma_i^b=f(\varepsilon_i^b)$ for a binder in the form of linear-reinforcing model (3.47). Substituting (3.47) into (3.65), we get:

$$E_f^e(1-\omega_f)=\frac{1}{k_F}[E_{averag.}(1-\omega_{averag.})$$

$$-(1-k_F)\frac{9K_bG_b(1-\omega_b)}{3K_b+G_b(1-\omega_b)}] \tag{3.66}$$

Substituting the values of K_b and G_b for a binder through E_b and v_b by formulae (3.49) into (3.66) and taking into account $E_b=\frac{9K_bG_b}{3K_b+G_b}$ and $\frac{G_b}{3K_b+G_b}=\frac{1}{3}(1-2v_b)$, we get:

$$E_f(1-\omega_f)=E_f^e\{1-\frac{1}{E_f-E_b(1-k_F)}[E_{averag.}\omega_{averag.}$$

$$-2E_b(1-k_F)\frac{(1+v_b)\omega_b}{3-(1-2v_b)\omega_b}]\} \tag{3.67}$$

whence, the longitudinal reinforcement function ω_f of a polymerized bundle from fibers will have the following form:

$$\omega_f=\frac{1}{E_f-E_b(1-k_F)}[E_{averag.}\omega_{averag.}$$

$$-2E_b(1-k_F)\frac{(1+v_b)\omega_b}{3-(1-2v_b)\omega_b}] \tag{3.68}$$

Thus, the longitudinal reinforcement function ω_f is determined with the help of data of first experimental test. From (3.68), it is seen that the necessary and sufficient condition for the case of linear-elastic deformability of a polymerized bundle from fibers with regard to arising interaction forces with an elastic matrix will be in form $\omega_f=0$, that is:

$$\omega_{averag.}=2\frac{E_b}{E_{averag.}}\frac{(1+v_b)\omega_b}{3-(1-2v_b)\omega_b}(1-k_F) \tag{3.69}$$

Hence, the following mechanical effect follows: In the case of elastic-deformability of binder, a composite, and a bundle from fibers regardless of influence of interaction forces with matrix subject to condition (3.69), a polymerized bundle from fibers with regard to interaction forces with a binder, may be deformed as a linear-elastic body.

Test 2: Tension with internal pressure. For defining the functions, $\Omega'_{\perp}$, $\Omega''_{\perp}$, $\nu'_{\perp} = \nu'^{e}_{\perp}(1-\Omega'_{\perp})$, and $\nu''_{\perp} = \nu''^{e}_{\perp}(1-\Omega''_{\perp})$ we use the results of experiments carried out on tubes with one longitudinally reinforced layer of bundles from fibers under joint action of the internal pressure $p_a = p$ and axial tension P. In this case, for $p_a = p$, $\sigma_3 = 0$ we shall have:

$$2\sigma^b_{\perp} = (\frac{R}{h} + \frac{1}{2})ap \tag{3.70}$$

Construct experimentally the dependence of change of cubic deformation θ on internal pressure p:

$$p = K^e_{averag.}(1-\Omega_{averag.})\theta \tag{3.71}$$

where, $K^e_{averag.}$ is an elastic volume modulus of a bundle from fibers and polymerized into a matrix, $\theta = \varepsilon_{ii} = \varepsilon_1 + 2\varepsilon_{\perp}$ is a cubic deformation. Use the formula of longitudinal tension of a tubular sample:

$$P = m(T_f + F_b\sigma^b_1) = mF_1E_{averag.}(1-\omega_{averag.})\varepsilon_1 \tag{3.72}$$

For establishing bonds between $\sigma^b_{\perp}$ and $\varepsilon^b_{\perp}$, we use experimental results represented in the form (3.71) and (3.72). For that, having substituted (3.71) in (3.70), take into account $\sigma^b_3 = 0$ and get a mathematical relation between $\sigma^b_{\perp}$ and $\varepsilon^b_{\perp}$ in the form:

$$\sigma^b_{\perp} = \frac{a}{2}K_{averag.}(1-\Omega_{averag.})(\varepsilon^b_1 + 2\varepsilon^b_{\perp}) \tag{3.73}$$

or

$$\varepsilon^b_{\perp} = \frac{1}{aK^e_{averag.}(1-\Omega_{averag.})}\sigma^b_{\perp} - \frac{1}{2}\varepsilon^b_1 \tag{3.74}$$

From (3.72) we have:

$$\frac{T_f}{F_1} = \frac{T_f}{F_f} k_F = E^e_{averag.}(1-\omega_{averag.})\varepsilon^b_1 - (1-k_F)\sigma^b_1 \qquad (3.75)$$

Substitute from (3.54), the expression σ^b_1 into (3.75) and take into account (3.73). So, we define the dependence of the stress of a polymerized bundle from fibers σ_f with regard to interaction forces with a binder, on longitudinal ε^b_1 and lateral $\varepsilon^b_\perp$ deformations, in the form:

$$\sigma_f = \frac{T_f}{F_f} = \frac{1}{k_F}\{\Big[E^e_{averag.}(1-\omega_{averag.}) - (1-k_F)\frac{9K_b\frac{\sigma_i}{\varepsilon_i}}{9K_b + \frac{\sigma_i}{\varepsilon_i}}$$

$$-\frac{a}{2}(1-k_F)\frac{9K_b - 2\frac{\sigma_i}{\varepsilon_i}}{9K_b + \frac{\sigma_i}{\varepsilon_i}}(1-\Omega_{averag.})K^e_{averag.}\Big]\varepsilon_1$$

$$-a(1-k_F)\frac{9K_b - 2\frac{\sigma_i}{\varepsilon_i}}{9K_b + \frac{\sigma_i}{\varepsilon_i}}K^e_{averag.}(1-\Omega_{averag.})\varepsilon_\perp\} \qquad (3.76)$$

In relations (3.76) and (3.58), comparing the coefficients for ε^b_1 and $\varepsilon^b_\perp$, we get an algebraic system with respect to $\nu^{'}_\perp$ and $\nu^{''}_\perp$:

$$\nu'_\perp = -\frac{1-k_F}{k_F}\frac{1}{E_f^e(1-\omega_f)}\frac{9K_b - 2\frac{\sigma_i}{\varepsilon_i}}{9K_b + \frac{\sigma_i}{\varepsilon_i}} aK^e_{averag.}(1-\Omega_{averag.})$$

$$-\frac{2}{k_F E_f^e(1-\omega_f)}\frac{9K_b + \frac{\sigma_i}{\varepsilon_i}}{9K_b - 2\frac{\sigma_i}{\varepsilon_i}}\{E^e_{averag.}(1-\omega_{averag.})$$

$$-\frac{1-k_F}{9K_b + \frac{\sigma_i}{\varepsilon_i}}[9K_b\frac{\sigma_i}{\varepsilon_i} + \frac{a}{2}(9K_b - 2\frac{\sigma_i}{\varepsilon_i})K^e_{averag.}(1-\Omega_{averag.})]$$

$$-k_F E_f^e(1-\omega_f)\}$$

$$\nu''_\perp = \frac{9}{k_F(9K_b - 2\frac{\sigma_i}{\varepsilon_i})}\{E^e_{averag.}(1-\omega_{averag.})$$

$$-\frac{1-k_F}{9K_b + \frac{\sigma_i}{\varepsilon_i}}[9K_b\frac{\sigma_i}{\varepsilon_i} + \frac{a}{2}(9K_b - 2\frac{\sigma_i}{\varepsilon_i})K^e_{averag.}(1 \tag{3.77}$$

$$-\Omega_{averag.})] - k_F E_f^e(1-\omega_f)\}$$

Here, $\nu'_\perp = \nu'^e_\perp(1-\Omega'_\perp)$ and $\nu''_\perp = \nu''^e_\perp(1-\Omega''_\perp)$; $\nu'^e_\perp$ and $\nu''^e_\perp$ are linear-elastic parts of lateral deformation of a polymerized bundle from fibers in the matrix, that are defined by formulae (3.29) and (3.37), and the reinforcement functions of a polymerized bundle in the matrix $\Omega'_\perp$ and $\Omega''_\perp$ defined from (3.77), extracting linear-elastic numbers from them.

Substitute (3.59) in (3.77), and by means of formulae (3.49), replace the elastic constants K_b and G_b by E_b and ν_b, and also take into account that

$$E_b = \frac{9K_bG_b}{3K_b + G_b}, \quad \frac{G_b}{3K_b + G_b} = \frac{1}{3}(1 - 2\nu_b)$$

In this case $\nu'_\perp$ and $\nu''_\perp$ will take the form:

$$\begin{aligned}
\nu'_\perp = &-\frac{1-k_F}{k_F}\frac{1}{E^e_f(1-\omega_f)}\frac{3K_b - 2G_b(1-\omega_b)}{3K_b + G_b(1-\omega_b)}aK^e_{averag.}(1-\Omega_{averag.})\\
&-\frac{2}{k_F E^e_f(1-\omega_f)}\frac{3K_b + G_b(1-\omega_b)}{3K_b - 2G_b(1-\omega_b)}\{E^e_{averag.}(1-\omega_{averag.})\\
&-\frac{1-k_F}{3K_b + G_b(1-\omega_b)}[9K_bG_b(1-\omega_b)\\
&+\frac{a}{2}(3K_b - 2G_b(1-\omega_b))\cdot K^e_{averag.}(1-\Omega_{averag.})] - k_F E^e_f(1-\omega_f)\}
\end{aligned}$$

$$\begin{aligned}
\nu''_\perp = &\frac{3}{k_F[3K_b - 2G_b(1-\omega_b)]}\{E^e_{averag.}(1-\omega_{averag.})\\
&-\frac{1-k_F}{3K_b + G_b(1-\omega_b)}[9K_bG_b(1-\omega_b)\\
&+\frac{a}{2}(3K_b - 2G_b(1-\omega_b))\cdot K^e_{averag.}(1-\Omega_{averag.})] - k_F E^e_f(1-\omega_f)\}
\end{aligned} \tag{3.78}$$

Extracting from (3.78) linear-elastic parts $\nu'^e_\perp$ and $\nu''^e_\perp$, define the reinforcement functions of a polymerized bundle from fibers $\Omega'_\perp$ and $\Omega''_\perp$ with regard to interaction forces with an elastic–plastic binder, in the form:

$$\Omega'_\perp = 1 - \frac{\nu'_\perp}{\nu'^e_\perp}, \quad \Omega''_\perp = 1 - \frac{\nu''_\perp}{\nu''^e_\perp} \tag{3.79}$$

Here $\nu'^e_\perp$ and $\nu''^e_\perp$ are of the form (3.37), $\nu'_\perp$ and $\nu''_\perp$ are of the form (3.78).

3.2.3 VISCO–ELASTIC PROPERTIES OF A BUNDLE FROM FIBERS WITH REGARD TO INTERACTION FORCES WITH A VISCO–ELASTIC MATRIX

It is known that polymer materials and fibers made up of them display essential dependence on time. Therefore, below we suggest a variant of a visco–elastic model of polymerized bundles from fibers (filaments and braids). Let a bundle from polymer materials that was polymerized into a polymer matrix be subjected to linear-visco elasticity. In the place of physico–mechanical model of a polymerized bundle from fibers with regard to interaction forces of a polymerized bundle from fibers with a polymer matrix, we suggest the following one (Aliyev, 1987, 1998):

$$\sigma_f(t)=\int_0^t \Pi_f(t-\tau)d\varepsilon_\ell^f(\tau)+v_\perp\int_0^t \Pi_\perp^f(t-\tau)d\varepsilon_\perp^b(\tau) \qquad (3.80)$$

or

$$\varepsilon_\ell(t)=\int_0^t J_f(t-\tau)d\sigma_\ell^f(\tau)-v_\perp'\int_0^t J_\perp^f(t-\tau)d\sigma_\perp^b(\tau) \qquad (3.81)$$

where, $II_f(t)=\sigma_\ell(t)/\varepsilon_\ell^0$ and $J_f(t)=\varepsilon_\ell(t)/\sigma_\ell^0$ are the relaxation and creeping functions of a polymerized bundle from fibers with regard to arising interaction forces between the polymerized bundle from fibers and a binder; $II_\perp^f(t)=\dfrac{\sigma_\perp^f(t)}{\varepsilon_\perp^0}$ and $J_\perp^f(t)=\dfrac{\varepsilon_\perp^f(t)}{\sigma_\perp^0}$ are the relaxation and creeping functions lateral to the polymerized bundle with regard to interaction forces on their contact boundary. For establishing relationship of the functions $II_\perp^f(t)$ and $J_\perp^f(t)$ from the properties of a polymer binder, bundles from fibers regardless of interaction forces and composite, we behave as follows. For a linear visco–elastic binder we have:

$$э_{ij}^{b}(t)=\int_{0}^{t}J_{b}(t-\tau)ds_{ij}^{b}(\tau),\quad \varepsilon^{b}=\frac{1}{3K_{b}}\sigma^{b}$$

or in the form:

$$s_{ij}^{b}(t)=\int_{0}^{t}П_{b}(t-\tau)d\,э_{ij}^{b}(\tau),\quad \sigma^{b}=3K_{b}\varepsilon^{b} \tag{3.82}$$

where, $s_{ij}^{b}=\sigma_{ij}^{b}-\sigma^{b}\delta_{ij}$, $э_{ij}^{b}=\varepsilon_{ij}^{b}-\varepsilon^{b}\delta_{ij}$, $3\sigma^{b}=\sigma_{ij}^{b}$; $J_{b}(t)=\frac{\varepsilon^{b}(t)}{\sigma_{0}^{b}}$ and $\ddot{I}_{b}(t)=\frac{\sigma^{b}(t)}{\varepsilon_{0}^{b}}$ are creeping and relaxation functions of the binder; $\tilde{A}_{b}(t)=2G_{0}^{b}\frac{dJ_{b}(t)}{dt}$ and $R_{b}(t)=-\frac{1}{2G_{0}^{b}}\frac{d\ddot{I}_{b}(t)}{dt}$ are the creeping and relaxation kernels; $J_{b}(0)=1/2G_{0}^{b}$, $R_{b}(0)=2G_{0}^{b}$. Applying the Laplace–Carson transform to (3.82), we get:

$$\bar{э}_{ij}^{b}=\bar{J}_{b}\bar{s}_{ij}^{b} \tag{3.83}$$

$$\bar{\varepsilon}^{b}=\frac{1}{3K_{b}}\bar{\sigma}^{b} \tag{3.84}$$

In principal stresses and deformations they will be in the form:

$$\begin{cases}\bar{\varepsilon}_{1}^{b}-\bar{\varepsilon}^{b}=\bar{J}_{b}(\bar{\sigma}_{1}^{b}-\bar{\sigma}^{b})\\ \bar{\varepsilon}_{2}^{b}-\bar{\varepsilon}^{b}=\bar{J}_{b}(\bar{\sigma}_{2}^{b}-\bar{\sigma}^{b})\\ \bar{\varepsilon}_{3}^{b}-\bar{\varepsilon}^{b}=\bar{J}_{b}(\bar{\sigma}_{3}^{b}-\bar{\sigma}^{b})\end{cases} \tag{3.85}$$

Allowing $2\varepsilon_\perp = \varepsilon_2 + \varepsilon_3$ and $2\sigma_\perp = \sigma_2 + \sigma_3$, from the first relation of (3.85), we get a plot expressed through the creeping function of the binder $\bar{J}_b$:

$$\bar{\sigma}_1^b = \frac{9K_b}{1+6K_b J_b}\left[\bar{\varepsilon}_1^b - \frac{2}{9K_b}(1-3K_b\bar{J}_b)\bar{\sigma}_\perp^b\right] \tag{3.86}$$

and also a plot expressed through the relaxation function of binder $\bar{\Pi}_b$ in the form:

$$\bar{\sigma}_1^b = \frac{9K_b\bar{\Pi}_b}{\bar{\Pi}_b + 6K_b}\left[\bar{\varepsilon}_1^b - \frac{2}{9K_b\bar{\Pi}_b}(\bar{\Pi}_b - 3K_b)\bar{\sigma}_\perp^b\right] \tag{3.87}$$

Putting together the second and third equations of (3.85), allowing (3.84), we get a relationship between the functions $\bar{\sigma}_\perp^b$ and $\bar{\varepsilon}_\perp^b$ in Laplace–Carson images by means of the creeping function $\bar{J}_b$ in the form:

$$9K_b\bar{\varepsilon}_\perp^b = (2+3K_b\bar{J}_b)\bar{\sigma}_\perp^b + (1-3K_b\bar{J}_b)\bar{\sigma}^b \tag{3.88}$$

From the relationship between the functions $\bar{\sigma}_\perp^b$ and $\bar{\varepsilon}_\perp^b$ for a binder in Laplace–Carson images by means of the relation function $\bar{\Pi}_c$ and taking into account the condition $\bar{J}_c\Pi_c = 1$, from (3.88), we get:

$$9K_b\bar{\varepsilon}_\perp^b = (2+3K_b J_b)\bar{\sigma} + (1-3K_b J_b)\bar{\sigma}^b \tag{3.89}$$

Substituting the value of $\bar{\sigma}_1^b$ from (3.86) into (3.88), establish relations between $\sigma_\perp^b$ and $\varepsilon_\perp^b$ in Laplace–Carson images in the form:

$$(1+6K_b\bar{J}_b)\bar{\varepsilon}_\perp^b = 3\bar{J}_b\bar{\sigma}_\perp^b + (1-3K_b\bar{J}_b)\bar{\varepsilon}_1^b \tag{3.90}$$

We get mathematical relation between $\sigma_{\perp}^{b}$ and $\varepsilon_{\perp}^{b}$ for a binder in Laplace–Carson image by means of the relation function $\Pi_b(t)$ from (3.2.90) allowing the condition $\bar{J}_b \Pi_b = 1$:

$$(\bar{\Pi}_b + 6K_b)\bar{\varepsilon}_{\perp}^{b} = 3\bar{\sigma}_{\perp}^{b} + (\bar{\Pi}_b - 3K_b)\varepsilon_1^{b} \tag{3.91}$$

Establish mathematical relation between the relaxation functions of the bundle $\Pi_f(t)$ and $\Pi_{\perp}^{f}(t)$, creeping functions $J_f(t)$ and $J_{\perp}^{f}(t)$. As (3.81) is the solution of integral equation (3.80), we apply the Laplace–Carson transform to equations (3.80) and (3.81):

$$\bar{\sigma}_f = \bar{\Pi}_f \bar{\varepsilon}_1^{b} + v_{\perp} \bar{\Pi}_{\perp}^{f} \bar{\varepsilon}_{\perp}^{b} \tag{3.92}$$

$$\bar{\varepsilon}_1^{b} = \bar{J}_f \bar{\sigma}_f + v_{\perp}' \bar{J}_{\perp}^{f} \bar{\sigma}_{\perp}^{b} \tag{3.93}$$

Substitute (3.92) and (3.90) into (3.93) and get:

$$\begin{aligned}&[(1 - \bar{J}_f \bar{\Pi} f)(1 + 6K_b \bar{J}_b) - v_{\perp} \bar{J}_f \bar{\Pi}_{\perp}^{f}(1 - 3K_b \bar{J}_b)]\bar{\varepsilon}_{\perp}^{b}\\ &= [3\bar{J}_b(1 - \bar{J}_f \bar{\Pi}_f) - v_{\perp}' \bar{J}_{\perp}^{f}(1 - 3K_b \bar{J}_b)\bar{\sigma}_{\perp}^{b}\end{aligned} \tag{3.94}$$

As (3.93) is the solution of equation (3.92), we equate the coefficients for $\varepsilon_{\perp}^{b}$ and $\sigma_{\perp}^{b}$ in equation (3.94) to zero and define the dependence of functions $\bar{J}_f$ and $v_{\perp} \bar{J}_{\perp}^{f}$ on given functions $\bar{\Pi}_f$ and $v_{\perp} \bar{\Pi}_{\perp}^{f}$ in the form:

$$\bar{J}_f = \frac{1 + 6k_b \bar{J}_b}{(1 + 6K_b \bar{J}_b)\bar{\Pi}_f (1 - 3K_b \bar{J}_b) v_{\perp} \bar{\Pi}_{\perp}^{f}}$$

$$v_{\perp}' \bar{J}_{\perp}^{f} = 3\bar{J}_b \frac{v_{\perp} \bar{\Pi}_{\perp}^{f}}{(1 + 6K_b \bar{J}_b)\bar{\Pi}_f + (1 - 3K_b \bar{J}_b) v \perp \bar{\Pi}_{\perp}^{f}} \tag{3.95}$$

Thus, defining experimentally the dependence $\Pi_f(t)\sim t$ and $\Pi_\perp(t)\sim t$ by formula (3.95), determine the Laplace–Carson image of longitudinal and lateral creeping functions of the polymerized bundle from fibers $\bar{J}_f(t)$ and $\bar{J}_\perp^f(t)$ with regard to arising interaction forces of a bundle with binder. By the inverse transformation of relation (3.95), it is easy to get the originals of the creeping functions $J_f(t)$ and $J_\perp^f(t)$. From (3.95), it is seen that for polymerized bundle-shaped bodies with regard to interaction forces with a visco–elastic matrix, the product of images of the creeping function $\bar{J}_f$ and relation function $\bar{\Pi}_f$ does not equal to unit as in the case of homogeneous visco–elasticity.

A way for experimental definition of visco–elastic functions $\Pi_f(t)\sim t$ and $\Pi_\perp(t)$ of a polymerized bundle from fibers with regard to interaction forces with visco–elastic binder

For defining the constant $\nu_\perp$ and the relaxation functions of the bundle $\Pi_b(t)$ and $\Pi_\perp(t)$ with regard to interaction forces with visco–elastic matrix it suffices to carry out two experiments.

Test 1: Definition of relaxation function of a polymerized bundle fibers $\Pi_f(t)$ under simple tension. In this case, the experiment is conducted provided $p_a = p_b = 0$ and $\sigma_\perp = 0$. We construct a diagram on a compositional pipe:

$$\Pi_{\text{averag.}}(t) = \frac{\sigma_\ell(t)}{\varepsilon_\ell^0} = f(t) \tag{3.96}$$

Then, the longitudinal force P applied to the end faces will equal:

$$P = m\left[\sigma_f(t)F_f + \sigma_1^b(t)F_b\right] = mF_1\int_0^t \Pi_{\text{averag.}}(t-\tau)d\varepsilon_\ell(\tau) \tag{3.97}$$

Substituting (3.80) and (3.82) in (3.97), we get:

$$\frac{P}{mF_1} = k_F \int_0^t \Pi_f(t-\tau)d\varepsilon_1(\tau) + (1-k_F)\int_0^t \Pi_b(t-\tau)d\varepsilon_1(\tau)$$
$$= \int_0^t \Pi_{averag.}(t-\tau)d\varepsilon_1(\tau) \quad (3.98)$$

and hence find the dependence of the relaxation function of a polymerized bundle from fibers on the relaxation function of the composite $\Pi_{averag.}(t)$ and binder $\Pi_b(t)$, in the form:

$$\Pi_f(t) = \frac{1}{K_F}\Pi_{averag.}(t) - \frac{1-k_F}{K_F}\Pi_b(t) \quad (3.99)$$

Hence, it is seen that if the relaxation function of the composite $\Pi_{averag.}(t)$ and relaxation function of the binder $\Pi_b(t)$ are known, then the relaxation function of a polymerized bundle from fibers $\Pi_b(t)$ allowing the interaction force with a polymer matrix is determined from (3.99). From (3.99) for $t = 0$, we have:

$$\Pi_f(0) = 2G_0^f = \frac{1}{k_F}\Pi_{averag.}(0) - \frac{1-k_F}{k_F}\Pi_b(0)$$
$$= \frac{2G^0_{averag.}}{k_f}\left[1-(1-k_F)G_b^0 / G^0_{averag.}\right] \quad (3.100)$$

that concides with the linear-elastic case. As $t \to \infty$, we get an expression of durable relaxation modulus of a polymerized bundle from fibers, in the form:

$$H_\sigma^f = \Pi_f(\infty) = \frac{1}{k_F}\Pi_b(\infty) - \frac{1-k_F}{k_F}\Pi_b(\infty)$$
$$= \frac{H^\sigma_{averag.}}{k_F}\left[1-(1-k_F)\frac{H_b^\sigma}{H^\sigma_{averag.}}\right] \quad (3.101)$$

where, $\Pi_{averag.}(\infty)=\sigma_{averag.}(\infty)/\varepsilon_0=H^{\sigma}_{averag.}$, $\Pi_b(\infty)=\sigma_b(\infty)/\varepsilon_0=H^{\sigma}_b$ are durable relaxation module of a composite and binder.

Test 2: Definition of lateral relaxation function of polymerized bundle from fibers $\Pi^f_{\perp}(t)$.

For defining the constants $v_{\perp}$ and the lateral relaxation function $\Pi^f_{\perp}(t)$, it is necessary to conduct an experiment on a reinforced tube under joint action of internal pressure and axial tensile force. During the test we measure the change of the volume relaxation function $V_{averag.}(t)=\frac{P(t)}{\theta_0}$ in time under constant internal volume and longitudinal deformation ε^0_{ℓ}. Then for $p_b=0$ and $\sigma_3=0$, we construct the dependence:

$$p(t)=\int_0^t V_{averag.}(t-\tau)d\theta(t) \tag{3.102}$$

where,

$$\theta=\varepsilon_{ii}=\varepsilon_1+2\varepsilon_{\perp} \tag{3.103}$$

In this case the applied longitudinal force equals:

$$\frac{P}{m}=\sigma_f(t)F_f+\sigma^b_1(t)F_b=F_1\int_0^t \Pi_{averag.}(t-\tau)d\varepsilon_1(\tau) \tag{3.104}$$

Establish a relation between $\sigma^b_{\perp}$ and $\varepsilon^b_{\perp}$ allowing the experimental results of (3.102) and (3.104). For that we substitute $\sigma_2=\left(\frac{R}{h}+\frac{1}{2}\right)p=ap$ in $\sigma^b_{\perp}=\frac{1}{2}\sigma_2$ and get:

$$\sigma^b_{\perp}=\frac{a}{2}p \tag{3.105}$$

Substitute (3.102) and (3.103) in (3.105) and get:

$$2\sigma_{\perp}^{b} = a\int_{0}^{t} V_{averag.}(t-\tau)d\theta(t) = a\int_{0}^{t} V_{averag.}(t-\tau)d\left(\varepsilon_{1} + 2\varepsilon_{\perp}^{b}\right) \quad (3.106)$$

In the Laplace–Carson image, there will be:

$$2\bar{\sigma}_{\perp}^{b} = a\bar{V}_{averag.}(\varepsilon_{1} + 2\varepsilon_{\perp}^{b}) \quad (3.107)$$

or

$$\bar{\varepsilon}_{\perp}^{b} = \frac{1}{a\bar{V}_{averag.}}\left(\bar{\sigma}_{\perp}^{b} - \frac{1}{2}\bar{\sigma}_{1}^{b}\right) \quad (3.108)$$

Applying the Laplace–Carson transform to (3.104), we get:

$$\bar{T}_{f} = F_{1}\bar{\Pi}_{averag.}\bar{\varepsilon}_{1} = \bar{\sigma}_{1}^{b}F_{b}^{/} \quad (3.109)$$

Substituting the value of $\bar{\sigma}_{1}^{b}$ from (3.87) and the value of $\bar{\sigma}_{\perp}^{b}$ from (3.107) to (3.109), we get the dependence of the stress of a polymerized bundle from fibers $\bar{\sigma}_{f}$ in Laplace–Carson images both on longitudinal deformation $\bar{\varepsilon}_{1}^{b}$ and on the mean value of the lateral stress $\bar{\varepsilon}_{\perp}^{b}$ in the binder in form:

$$\bar{T}_{f}/F_{1} = \bar{\sigma}_{f} = \{\bar{\Pi}_{averag.} - \frac{1-k_{F}}{6K_{b} + \bar{\Pi}_{b}}[9K_{b}\bar{\Pi}_{b} - (\bar{\Pi}_{b} - 3K_{b})a\bar{V}_{averag.}]\}\bar{\varepsilon}_{1} + 2a\bar{V}_{averag.}(1 - \frac{\bar{\Pi}_{b} - 3K_{b}}{6K_{b} + \bar{\Pi}_{b}})\bar{\varepsilon}_{\perp}^{b} \quad (3.110)$$

Substituting in (3.109) the value of $\bar{\sigma}_{1}^{b}$ from (3.87), we get the dependence of stress of a polymerized bundle from fibers $\bar{\sigma}_{f}$ in Laplace–Carson images both on longitudinal deformation $\bar{\varepsilon}_{1}^{b}$ and on the mean value of the lateral stress $\bar{\sigma}_{\perp}^{b}$ in the binder, in form:

$$\bar{T}_f / F_1 = \bar{\sigma}_f = [\bar{\Pi}_{averag.} - (1-k_F)\frac{9K_b\bar{\Pi}_b}{6K_b+\bar{\Pi}_b}]\bar{\varepsilon}_1$$

$$+2(1-k_F)\frac{\bar{\Pi}_b - 3K_b}{6K_b+\bar{\Pi}_b})\bar{\sigma}_\perp^b \qquad (3.111)$$

Comparing the coefficients for (3.92) and (3.110), define the image of the relaxation function of a polymerized bundle from the fibers $\bar{\Pi}_f$ and $\nu_\perp \bar{\Pi}_\perp^f$ in the form:

$$\left\{ \begin{aligned} \bar{\Pi}_f &= \bar{\Pi}_{averag.} - \frac{1-K_F}{6K_b+\bar{\Pi}_b}[9K_b\bar{\Pi}_b - (\bar{\Pi}_b - 3K_b)a\bar{V}_{averag.}] \\ \nu_\perp \bar{\Pi}_\perp^f &= 2a\bar{V}_{averag.} 1 - \frac{\bar{\Pi}_b - 3K_b}{6K_b+\bar{\Pi}_b} \end{aligned} \right\} \qquad (3.112)$$

Now, substitute (3.112) in (3.95) and define the dependence of the creeping function of a polymerized bundle from the fibers $\bar{J}_f$ and $\nu_\perp \bar{J}_\perp^f$ in Laplace–Carson images on the experimental functions $\bar{\Pi}_{averag.}$ and $\bar{V}_{averag.}$ and also on the relaxation functions of the bundle $\bar{\Pi}_f$ and $\Pi_\perp^f$, in the form:

$$\{\bar{J}_f = \frac{\bar{\Pi}_b + 6K_b}{(\bar{\Pi}_b + 16K_b) - (1-K_F)\{9K_b\bar{\Pi}_b - (\bar{\Pi}_b - 3K_b)a\bar{V}_{averag.}[1+2(\bar{\Pi}_b - 3K_b)]\}}$$

$$\nu_\perp \bar{J}_\perp^f = \frac{6(1-K_F)(\bar{\Pi}_c - 3K_c)1}{(\bar{\Pi}_b + 6K_b)\left\{(\bar{\Pi}_b + 6K_b)\bar{\Pi}_{averag.} - (1-K_F)\{9K_b\bar{\Pi}_b - a\bar{V}_{averag.}(\bar{\Pi}_b - 3K_b)[1+2(1+2\bar{\Pi}_b - 3K_b)]\}\right\}} \qquad (3.113)$$

Models (3.80) and (3.81) are assumed to be completely defined if experimental functions (3.112) or (3.113) are known.

3.2.4 ON AN APPROXIMATE VISCO–ELASTIC MODEL OF A POLYMERIZED BUNDLE FROM FIBERS WITH REGARD TO INTERACTION FORCES WITH VISCOUS–ELASTIC MATRIX

For the bundles with total impregnation of the binder of all fibers and strong arrangement of bundles in the composite; for the case when the lateral stress quantities weakly manifest themselves, that is, they are small;

when the stress–strain state of the bundle is independent of the form of the applied load, on the other hand for reducing the amount of experimental works we suggest the following approximate physico–mechanical model of a polymerized bundle from fibers with regard to arising interaction forces with a visco–elastic matrix. For this case, we suggest the deformation laws of a bundle from fibers situated in a viscous–elastic matrix both in the variant with coupling and in the case without influence of a polymer medium, in the following form (Aliyev, 1987):

$$\sigma_f = {}^{T_f}\!\!/_{F_f} = E_f^{longterm}\{\varepsilon\ell + v_\perp \varepsilon_\perp$$
$$-\int_0^t R_0(t-\tau)\varepsilon_\ell(\tau)d\tau - v_\perp \int_0^t \Gamma_0(t-\tau)\varepsilon_\perp(\tau)d\tau] \quad (3.114)$$

$$\sigma_f^0 = {}^{T_f^0}\!\!/_{F_f} = E_f^0[\varepsilon_\ell - \int_0^t R_0(t-\tau)\varepsilon_\ell(\tau)d\tau] \quad (3.115)$$

and the binding material is subjected to the law:

$$\sigma^b = E_b[\varepsilon_\ell^b - \int_0^t {}_0(t-1)\varepsilon_\ell^b(\tau)d\tau] \quad (3.116)$$

It is assumed that the lateral contraction of a polymerized bundle from fibers is proportional to the longitudinal deformation of a binder, that is,

$$\varepsilon_\perp^b = v_b \varepsilon_\ell^b \quad (3.117)$$

It is also assumed that the ratio of strength of a polymerized bundle from fibers σ_{fs} with regard to interaction forces with a polymer matrix to the strength of a bundle without medium σ_{fs} is proportional to the ratio of their long term module $E_{fo}^{longterm}$ and $E_{fo}^{longterm}$:

$$\frac{\sigma_{fs}}{\sigma_{fs}^{c}} = \frac{E_f^{longterm}}{E_{f0}^{longterm}} = k\sigma \tag{3.118}$$

where,

$$E_{f0}^{longterm} = E_{f0}^{longterm}[1 - \int_0^\infty R(\xi)d\xi],$$

$$E_f^{longterm} = E_f[1 - \int_0^\infty R_f(\xi)d\xi] \tag{3.119}$$

In this case, the total longitudinal force applied to the cross section of the sample in the form of a bar will be composed of forces applied to the polymerized bundle and applied to the binding material, in the form:

$$P = nT_f + (EF)_b[\varepsilon_\ell - \int_0^t \Gamma_0(t-\tau)\varepsilon_\ell(\tau)d\tau = (EF)_{averag.}[\varepsilon_\ell - \int_0^t V_{averag.}(t-\tau)\varepsilon_\ell(\tau)d\tau] \tag{3.120}$$

Here, $E_{averag.}$ and $V_{averag.}(t)$ represent an elasticity modulus and a relaxation kernel respectively, of an averaged composite material. Substituting (3.114)–(3.118) in (3.120), we define the coefficient of a lateral contraction of a polymerized bundle from fibers, in the form:

$$v_\perp = \frac{1}{v_b[\varepsilon_\ell - \int_0^t \Gamma_0(t-\tau)\varepsilon_\ell(\tau)d\tau]}$$

$$-\frac{E_{averag.} - E_b(1-k_F n)}{nk_F k_\sigma E_{f0}^{longterm}}]\varepsilon_\ell - \int_0^t R_0(t-\tau)\varepsilon_\ell(\tau)d\tau +$$

$$+\frac{1}{nk_F k_\sigma E_{f0}^{longterm}}[E_{averag.}\int_0^t V_{averag.}(t-\tau)\varepsilon_\ell(\tau)d\tau$$

$$-E_b(1-k_F n)\int_0^t \Gamma_0(t-\tau)\varepsilon_\ell(\tau)d\tau]\} \qquad (3.121)$$

Introduce the denotation:

$$E_{b0}^{longterm} = E_b^0[1-\int_0^\infty \Gamma_0(\xi)d\xi] = \lambda_2 E_{f0}^{longterm}$$

is the long term modulus of the binder,

$$E_{averag.}^{longterm} = E_{averag}^0[1-\int_0^\infty V_{averag.}(\xi)d\xi] = \lambda_1 E_{f0}^{longterm}$$

is the bundle long term module of the composite,

$$\lambda_b = \frac{E_b^{longterm}}{E_b^0}, \quad \lambda_f = \frac{E_{0f}^{longterm}}{E_{0f}^0} \qquad (3.122)$$

Then (3.121) will take the following compact form:

$$\nu_\perp = \frac{1}{\nu_b \lambda_b}\{\lambda_f - \frac{1}{nk_F k_\sigma E_{0f}^{longterm}}[E_{averag.}^{longterm} - (1-nk_F)E_b^{longterm}]\} \qquad (3.123)$$

Thus, the suggested model (3.114) becomes completely defined if the constants $E_f^{longterm}$ and $\nu_\perp$ experimentally defined from formulae (3.118) and (3.123) that in their turn are defined from one dimensional experiment on a polymerized bundle from fibers situated in a polymer matrix, are known.

Special case 1: If the deformability of the bundle from fibers regardless of influence of a medium is subjected to the linear-elastic one, that is, $\lambda_b = 1$, $\lambda_2 = E_b^0 / E_{0f}^{longterm}$, then we shall have the following simplified model of a polymerized bundle:

$$\sigma_f = T_f / F_f = E_f^{lonterm} \{\varepsilon_\ell - \nu_\perp \varepsilon_\perp - \int_0^t R_0(t-\tau)\varepsilon_\ell(\tau)d\tau ,$$

$$k_\sigma = \sigma / \sigma_{ns}^0 \qquad (3.124)$$

$$E_f^{longterm} = k_\sigma E_{0f}^{longterm} ,$$

$$\nu_\perp = \frac{1}{\nu_b}\{\lambda_f - \frac{1}{nk_f k_\sigma}[\lambda_1 - (1 - Nk_F)\lambda_1]\} \qquad (3.125)$$

Special case 2: If a bundle is linear-elastic, and a polymer matrix is linear-viscous–elastic, that is, $\lambda_f = 1$, $\lambda_1 = E_{averag.}^{longterm} / E_{of}$, $\lambda_2 = E_b^{longterm} / E_{of}$, then we shall have a model of a polymerized bundle from fibers in the form:

$$\sigma_f = E_f[\varepsilon_\ell + \nu_\perp \varepsilon_\perp - \int_0^t \Gamma_0(t-\tau)\varepsilon_\perp(\tau)d\tau] \qquad (3.126)$$

where, $E_f = k_\sigma / E_{0f}$,

$$\nu_\perp = \frac{1}{\nu_b \lambda_b}\{1 - \frac{1}{nk_F k_\sigma}[\lambda_1 - (1 - nk_F) \cdot E_0^{longterm} / E_{of}]\} \qquad (3.127)$$

Special case 3: If the bundle and polymer matrix are subjected to linear elastic laws, then the model of a polymerized bundle from fibers will be:

$$\sigma_f = E_f(\varepsilon_\ell + \nu_\perp \varepsilon_\perp)$$

$$\lambda_b = \lambda_f = 1, \quad \lambda_1 = E^0_{averag.} / E_{of}, \quad \lambda_2 = E^0_b / E_{of} \tag{3.128}$$

$$\nu_\perp = \frac{1}{\nu_b}[1 - \frac{E_{averag.} - (1 - nk_F)E^0_b}{nk_F k_\sigma E_{of}}] \tag{3.129}$$

3.3 EXPERIMENTAL INVESTIGATIONS OF MECHANICAL PROPERTIES OF BUNDLES MADE UP OF GLASS FIBERS IN QUASISTATIC INTERACTION WITH A POLYMER MATRIX

It is known that mechanical properties (elasticity modulus E, ultimate strength σ_a, and the curve $\sigma \sim \varepsilon$) of elementary fibers are independent of their functioning conditions, that is, in the construction's body they possess the same mechanical properties that free fibers tested in air. But, the mechanical properties of polymerized bundles from fibers (filaments) depend on stress state of a binding material in which they are reinforced, generally speaking, they do not coincide with the properties of the bundle while testing them, regardless of influence of binding medium. In this connection, in Section 3.2.1–3.2.4, we suggested the models of mechanical deformation of polymerized bundles from fibers with a matrix and give their mathematical ground (Aliyev, 1987).

In this section we give experimental ground to the above given theoretical models of deformation of polymerized bundle-shaped bodies; show influence of the ambient binding medium on physico–chemical properties of polymerized bundles from fibers (filaments). The glass braids and glass filaments of different brands were taken as the tested material.

The glass braids of the brands BS6-34 × 1 × 3 × n and BS6-36 × 1 × 3 × n were experimentally investigated. Such glass filaments consist of 400 elementary fibers of the text BS6-34 or BS6-36 twisted in 70–80 $\text{revolutions}/_{\text{min}}$; the glass braids are the fiber filaments folded 5, 10, and 15 times and twisted in opposite direction in 70–80 $\text{revolutions}/_{\text{min}}$. Definition of mechanical properties of glass fibers folded 5, 10, and 15 times (elasticity modulus E, ultimate strength $\sigma_{\text{â}}$, and the curve $\sigma \sim \varepsilon$) was realized in tearing machine. The quantity of the applied load was taken by means of loads scale and elongation of the sample was measured by a measuring rule. The diagram "stress deformation" for glass braids folded 5, 10, and 15 times (Figure 3.2) were constructed by experimental data. The elasticity modulus E_f and ultimate strength $\sigma_{\text{â}}$ of glass braids were determined regardless of influence of binding medium (Aliyev, 1987):

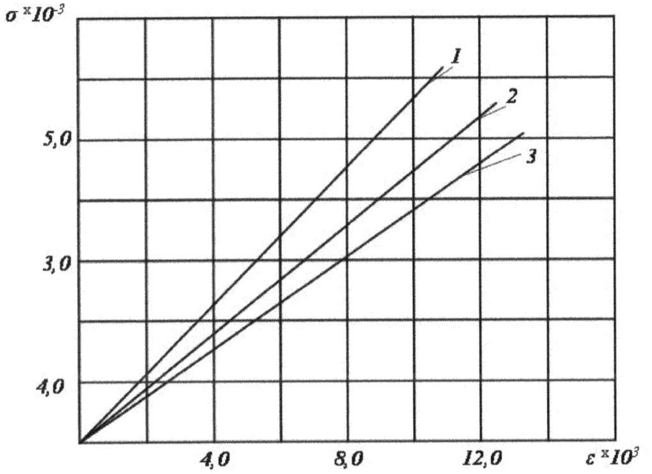

FIGURE 3.2 Diagrams of stress-deformation $\sigma = f(\varepsilon)$ for a glass bundle from fibers of the brand BS6-34 × 1 × 3n (1, 2, and 3 corresponds to n = 5, 10, 15 fold of the glass bundle, respectively)

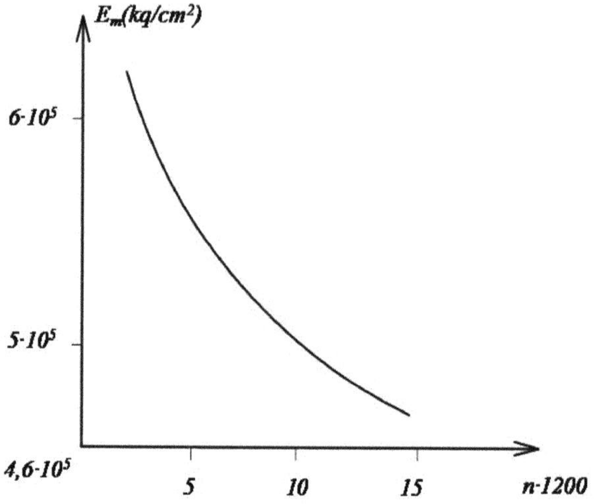

FIGURE 3.3 Dependence of elasticity modulus of glass bundle on fibers E_n of the brand BS6-34 × 1 × 3 × 3 n on the quantity content of elementary fibers in it

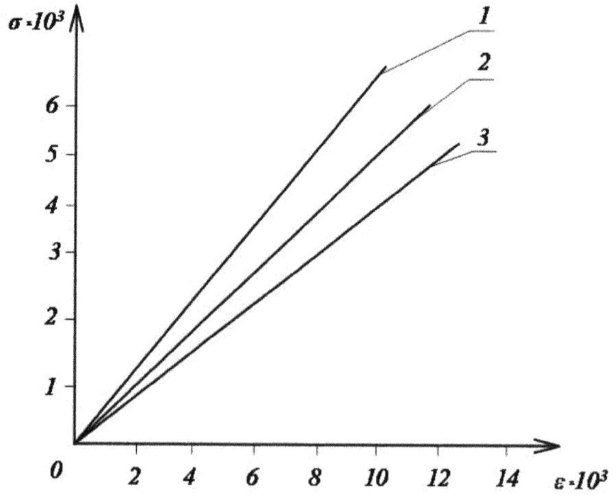

FIGURE 3.4 Diagram of stress-deformation $\sigma = f(\varepsilon)$ for a damaged glass bundle from the fibers of them BS6-34 × 1 × 3n (1, 2, 3 corresponds to n = 5, 10, 15 fold of the glass bundle, respectively)

$$E_5 = 0.595 \cdot 10^6 \ \frac{\text{kgf}}{\text{cm}^2}; \ \sigma_{\hat{a}5} = 8{,}423.1 \ \frac{\text{kgf}}{\text{cm}^2}$$

$$E_{10} = 0.521 \cdot 10^6 \ \frac{\text{kgf}}{\text{cm}^2}; \ \sigma_{\hat{a}10} = 8{,}389.3 \ \frac{\text{kgf}}{\text{cm}^2}$$

$$E_{15} = 0.475 \cdot 10^6 \ \frac{\text{kgf}}{\text{cm}^2}; \ \sigma_{\hat{a}5} = 8{,}016.3 \ \frac{\text{kgf}}{\text{cm}^2}$$

Hence it is seen that the elasticity modulus E_f and the ultimate strength σ_{ef} of glass braids decrease according to increase of the amount of elementary fibers in them (Figure 3.3). The influence of manufacturing methods of glass rubber pipes on mechanical properties of glass braids, that is, process of splicing, braiding, and coiling in roping, braiding, and coiling plants was also studied. The glass braids of the brand BS6-34 × 1 × 3 × n were tested. The diagrams "stress-deformation" were constructed for damaged glass bundles in 5, 10, and 15 folds (Figure 3.4), the elasticity modulus E_f and ultimate strength σ_{ef} were determined

$$E_5 = 0.524 \cdot 10^6 \ \frac{\text{kgf}}{\text{cm}^2}; \ \sigma_{\hat{a}5} = 6{,}751 \ \frac{\text{kgf}}{\text{cm}^2}$$

$$E_{10} = 0.442 \cdot 10^6 \ \frac{\text{kgf}}{\text{cm}^2}; \ \sigma_{\hat{a}10} = 6{,}451 \ \frac{\text{kgf}}{\text{cm}^2}$$

$$E_{15} = 0.427 \cdot 10^6 \ \frac{\text{kgf}}{\text{cm}^2}; \ \sigma_{\hat{a}5} = 6{,}268 \ \frac{\text{kgf}}{\text{cm}^2}$$

The results of experimental investigations show that technological processes of manufacturing of articles, in particular, of glass rubber pipes reduce to damaging of bundles from fibers and this lowers their mechanical properties (E_f by 15% and σ_{ef} by 30%).

The influence of the kind of stress state of a binding material (of a rubber) on mechanical properties of bundles from fibers with the given structure, in which they were reinforced, is also studied. We tested the thick walled stripes with the area of cross section 1.33cm^2, along which the braids made up of fibers in the amount of 1, 3, and 5 folds were reinforced. Change of mechanical properties of bundles from glass braids (elasticity modulus E_f, averaged coefficient of lateral compression of the bundle from fibers in the medium ν_f, and ultimate strength σ_{ef}) with regard to influence of the kind of stress state of a binding material (of a rubber) were determined as follows.

Now let the bundles from glass fibers situated in the binding medium with coupling and without influence of a binding medium be subjected to the laws:

$$T_f = (EF)_f(\varepsilon_\ell + \nu_f \varepsilon_\perp) \tag{3.130}$$

$$T_0 = E_0 F_0 \varepsilon_\ell \tag{3.131}$$

where E_f, ν_f, F_f are elasticity modulus, averaged coefficient of lateral compression of a bundle and cross section area of a bundle situated in a binding medium, ε_ℓ and $\varepsilon_\perp$ are longitudinal and lateral deformations of the bundle, E_0 is an elasticity modulus of the bundle without influence of a binding medium.

It is assumed that lateral contraction of a bundle from fibers in the medium nearly depends on longitudinal deformation of the binding material:

$$\varepsilon = -\nu_c \varepsilon_\ell \tag{3.132}$$

and the ultimate strength of a bundle from fibers in a binding medium and regardless of influence of the medium are directly proportional to their module:

$$\frac{\sigma_f^{in\,medium}}{\sigma_f^{free}} = \frac{E_f}{Å_{0f}} = \kappa_\sigma \tag{3.133}$$

Substituting (3.131) and (3.133) in (3.130) with regard to the force formula applied to the cross section of the sample:

$$P = nT_f + (EF)_b \varepsilon_\ell = E_{averag.} \cdot F \cdot \varepsilon_\ell \tag{3.134}$$

Define the elasticity modulus E_f and averaged coefficient ν_f of the lateral compression of the bundle from fibers with regard to influence of stress state of a binding medium, in the form:

$$E_f = k_\sigma E_0 \tag{3.135}$$

$$\nu_f = \frac{1}{\nu_b}[1 - \frac{1}{n}\frac{E_{averag.} - E_b(1 - nk_F)}{k_F k_\sigma E_0}] \tag{3.136}$$

Here, F is the cross section area of the sample, $E_{averag.}, E_b$ are elasticity module of a composite sample and a binding material, n is the amount of bundles in the sample.

TABLE 3.1 Physico–mechanical properties of some types of bundles from fibers

Brand of the glass bundle	**BS6-34 × 1 × 3 × 5**	**BS6-34 × 1 × 3 × 10**	**BS6-34 × 1 × 3 × 15**
Cross-section area of the glass bundle $F_f(cm^2)$	$237 \cdot 10^{-5}$	$465 \cdot 10^{-5}$	$686 \cdot 10^{-5}$
Elasticity modulus of a Bundle from glass fiber $E_0(^{kgf}/_{cm^2})$	$0.524 \cdot 10^6$	$0.442 \cdot 10^6$	$0.427 \cdot 10^6$

TABLE 3.1 *(Continued)*

Ultimate strength of a glass bundle $\sigma_{aB}\left({}^{kgf}/_{cm^2}\right)$	6,751	6,451	6,268
Cross section area of the sample F (cm^2)	1.33	1.33	1.33
$k_F = {}^{F_f}/_F$	175·10-5	350·10-5	516·10-5
Ultimate strength of the glass bundle in rubber $\sigma_{fB}\left({}^{kgf}/_{cm^2}\right)$	8,439	8,172	8,105
Elasticity modulus of a bundle from fibers in rubber $E_f\left({}^{kgf}/_{cm^2}\right)$	0.655·106	0.561·106	0.551·106
Averaged coefficient of lateral compression of the glass bundle ν_f	0.684	0.585	0.512

3.4 PHYSICO–MECHANICAL DEFORMATION MODEL OF A POLYMERIZED BUNDLE FROM FIBERS IN THE NECK ZONE

Adhesive strength of a composite material is connected with the interaction force effect in the contact area between the reinforcing force fibrous structure and a binder. The binding medium itself also exerts an essential effect on the character of mechanical deformation of a fibrous structure situated in the binding medium. Therefore, it is necessary to know the models of mechanical deformation of a polymerized bundle-shaped body with regard to interaction forces with a polymer matrix. Especially, if a polymerized bundles from fibers has a necking (Aliyev 1987, 2011, 2012).

The necking phenomenon arising in structural elements from visco–plastic, elastico–plastic or visco–elastic materials is in the following. The prime extension test on samples made of plastic metals show that beginning with some moment, the deformation state of the smooth part of the sample stops to be homogeneous. Therewith, the arising deformation is concentrated in a small area of the sample. There happens local contraction, and elongation of this area. This contracted zone is called a neck. The following questions are interesting from scientific point of view: the first question: in what form will be physico–mechanical dependence between stress and deformation at the neck zone points; the second question: in what form will be stress–strain state in the area of neck zone points. Till today, the scientific investigations were carried out on the second theme, that is, on studying stress–strain state of structural elements at the necking zone points. These investigations were carried out at the assuming initial period of necking. In other words, at disappearance of homogeneity, the mechanical deformation model of the material is accepted as before necking. A. A. Ilyushin (Ilyushin, 1940, 2011) first has investigated the neck phenomenon in such a statement. He has first given the statement of the problem on stability of visco–elastic flow a strip and a rod; composed differential equations and boundary conditions of the problem for defining visco–plastic and plane parallel flow. The complexity of this problem was connected with Lagrange method for describing the motion of continuous medium. A. Yu. Ishlinsky (Ishlinskiy, 1943), used the method based on Euler's way for describing the medium's motion, and investigated the visco–plastic flow of a stripe and a rod. Joukov A. M. (Jukov, 1949) considered a problem on stress–strain state at the neck zone of a strip and a rod made of elastic–plastic material. He numerically compared theoretical results with experimental ones. Further, a number of authors carried out expanded experimental investigations on finding real stresses at the neck zone, that is the stresses with regard to cross section area at the neck zone, and also analyzed the character of the arising deformational in homogeneities of materials (Aliyev, 2011, 2012).

In 1970–2000, the necking phenomenon got a new scientific interest. It was connected with appearance of sandwich and composite materials executed on the base of polymers. One of the important problems in the strength of sandwich and composite materials is to provide adhesive strength between the layers of the material. And this directly depends on the necking character at the contact zone of heterogeneous materials. A number of experimental investigations on studying the interlayer failure in sandwich and composite materials were carried out. The fundamental result of these investigations is the creation of some criteria of adhesive strength in sandwich and composite materials (Aliyev, 1987). So, in (Aliyev, 1987) a general conception of mechanics of composite materials with regard to arising interaction forces between the elements of the composite is created. In this work, a revised conjecture of adhesive bonds forces between the layers is suggested. On these bases, a new theory of strength stability and vibration of sandwich-reinforced flexible thick-walled pipes and thin-walled shells with regard to arising interacting forces between the elements of the composite (Aliyev, 1987, 1995, 1998, 2012, 2012) is created.

However, the problem on definition of mechanism of origination of deformational heterogeneity at the neck zone, and also establishment of physico–mechanical dependence between stress and deformation at the neck zone points, has not been solved till today because of its complexity. Till present day, this problem belongs to the class of unsolved problems of continuum mechanics.

The mechanism of origination of deformational heterogeneity at the neck zone of a polymerized bundle from fibers, and also representation of a specific physico–mechanical model that establishes dependence between stress and deformation at the neck zone points was first suggested in (Aliyev, 2011, 2012). An experimental-theoretical method on the base of which was established that the character of deformational heterogeneity at the neck zone of a polymerized bundle from fibers is linear with respect to the longitudinal coordinate, was suggested in this work.

In the present section, we suggest an experimental-theoretical method that allows establishing the character of deformational heterogeneity at the neck zone of a polymerized bundle from fibers. In particular, mathematical dependence of mechanical properties of the material of polymerized bundles from fibers on longitudinal coordinate of the sample is established.

The suggested physico–mechanical model of deformation of a polymerized bundle from fibers at the neck zone with a binder is constructed under the following suppositions:

the cross section area of a bar made of polymerized fibers at the neck zone is a function of the longitudinal coordinate x. The cross section area has the form of the area of a circle the neck contour equation is assumed to be known, the thickness of a polymerized bundle from fibers is such that radial stress σ_r in the interval $|r| \leq r_{bundle}$ changes inconsiderably.

Under such assumptions, the normal $\vec{\varepsilon}_\nu$ and tangential $\vec{\varepsilon}_\tau$ deformation vectors arising on the boundary of lateral surface of a polymerized bar with a binder will characterize the deformed state of each internal point of a polymerized bar made of fibers. On the other hand the deformation vectors $\vec{\varepsilon}_\nu$ and $\vec{\varepsilon}_\tau$ at the points of the boundary of a polymerized bar with a binder will create corresponding stress vectors $\vec{\sigma}_\nu = E_f(x)\lambda_\nu(x)\vec{\varepsilon}_\nu$ and $\vec{\sigma}_\tau = E_f(x)\lambda_\tau(x)\vec{\varepsilon}_\tau$ that will be the functions of the form of neck contour.

Here, the deformation vector $\vec{\varepsilon}$ at the points of lateral surface of the neck is expressed by the normal $\vec{\nu}$ and tangential $\vec{\tau}$ unit vectors and also by Cartesian coordinate $(\vec{i},\vec{j},\vec{k})$ in the following form (Ilyushin and Lenskiy, 1959):

$$\vec{\varepsilon} = \vec{\varepsilon}_\nu + \vec{\varepsilon}_\tau = \varepsilon_\nu\vec{\nu} + \varepsilon_\tau\vec{\tau} \tag{3.137}$$

$$\vec{\varepsilon} = \vec{\varepsilon}_x\ell + \vec{\varepsilon}_y m + \vec{\varepsilon}_z n \tag{3.138}$$

where,

$$\varepsilon_\nu = \vec{\varepsilon}\vec{\nu}, \quad \varepsilon_\tau = \vec{\varepsilon}\vec{\tau} \tag{3.139}$$

$$\vec{\nu} = \vec{i}Cos\alpha + \vec{j}Sin\alpha = -\vec{i}\frac{dy}{ds} + \vec{j}\frac{dx}{ds},$$

$$\vec{\tau} = -\vec{i}Sin\alpha + \vec{j}Cos\alpha = -\vec{i}\frac{dx}{ds} - \vec{j}\frac{dy}{ds} \tag{3.140}$$

and the vectors $\vec{\varepsilon}_x$, $\vec{\varepsilon}_y$, $\vec{\varepsilon}_z$ are determined by the known way through the Cartesian coordinates ε_{ij} [22].

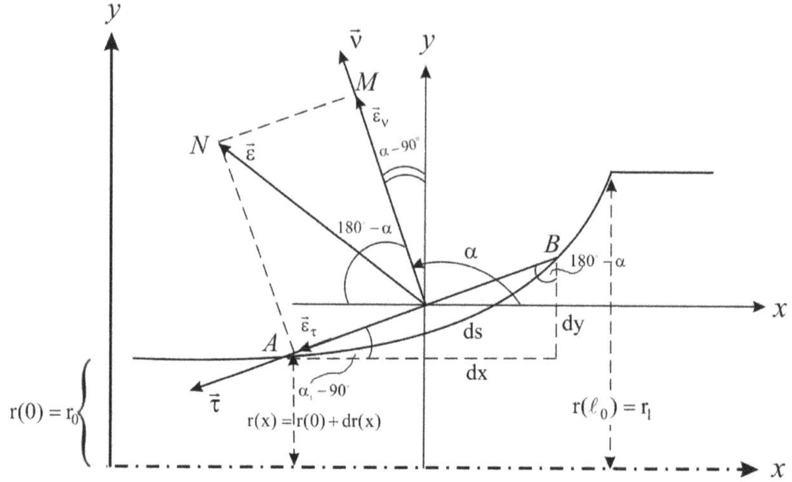

FIGURE 3.5 Picture of deformation of polymerized bundle from fibers at the neck zone

From the equilibrium equation of forces per unit element dx at the neck zone with the lateral annular cross section area $dF(x)$ and lateral surface of the from of a hyperboloid of one sheet $dL(x)$ we suggest the following dependence of the stress vector $\vec{\sigma}_f(x)$ on longitudinal deformation vector $\vec{\varepsilon}_{\delta} = \varepsilon_{\delta}\vec{i}$ and deformation vectors $\vec{\varepsilon}_v$ and $\vec{\varepsilon}_\tau$, acting on a lateral surface of a polymerized bar at the neck zone, in the form (Figure 3.5):

$$\vec{\sigma}_f(x) = E_f(x)[\vec{\varepsilon}_x + \lambda_v(x)\vec{\varepsilon}_v + \lambda_\tau(x)\vec{\varepsilon}_\tau] \tag{3.141}$$

Here, $E_f(x)$, $\lambda_v(x)$, $\lambda_\tau(x)$ are mechanical characteristics of a polymerized bar made of fibers at the neck's cross section area. They will depend on longitudinal coordinate x of the bar. These mechanical characteristics will depend on the ratio of elementary area of lateral surface $dL(x)$ to the cross section area $dF(x)$ of a polymerized bar in the form $f(x) = \frac{dL(x)}{dF(x)} = \frac{\ell}{2\eta}\frac{1}{x}$, and also on the parameter $\eta = \frac{\delta}{\ell}$, that is the ratio of the neck's depth δ to its length ℓ. The above stated mechanical characteristics will be determined by the following form special experiments.

3.4.1 EXPERIMENTAL-THEORETICAL METHOD FOR DEFINING MECHANICAL CHARACTERISTICS OF A POLYMERIZED BAR MADE OF FIBERS $E_f(x)$, $\lambda_v(x)$, $\lambda_\tau(x)$ AND THE NECK ZONE

Let's consider bar a polymerized into a matrix and made on the base of a fibrous structure, situated under the action of a longitudinally applied tensile force P. And while stretching, the polymerized bar is necking (Figure 3.5). Locate the system of coordinates in the narrow place of the neck with the following geometrical characteristics: $x = 0$ the neck of length $2\ell_0$ has the depth δ_0. It is assumed that the form of the neck's contour is known and is a circle. Assume that at the points $x = 0$ and $x = \ell_0$ of the neck, the values of longitudinal elasticity module of a polymerized bar $E_f(0)$ and $E_f(\ell_0)$ are known. In this case, we represent the dependence of the elasticity modulus $E_f(x)$ at any x cross section of the neck of a polymerized bar made of fibers on the longitudinal coordinate x in the following form:

$$E_f(x) = E_f(\ell_0)[1 + (\frac{E_f(0)}{E_f(\ell_0)} - 1)(1 - \frac{x}{\ell_0})] \text{ for } 0 \le x \le \ell_0 \quad (3.142)$$

Here, $E_f(\ell_0)$ is the value of elasticity modulus of a polymerized bar at the section $x = \ell_0$ of the neck, that is at the neck's end. The value of the elasticity modulus at the neck's end $E_f(\ell_0)$ will coincide with the value of elasticity modulus of a polymerized bar out of the neck, whose definition method was suggested in (Aliyev, 1987, 2012), and equals:

$$E_f(\ell_0) = k_\sigma E_0 \quad (3.143)$$

where E_0 is an elasticity modulus of the bundle without influence of a binding medium.

By the smallness of the influence of lateral deformation of a binding material at the very narrow part of the neck $x = 0$ of the bar on lateral deformation of a polymerized bar, we can assume that at the narrow part of the neck $x = 0$ only the fibers directed strongly in the longitudinal direction work. Therefore, in place an elasticity modulus $E_f(0)$ at the cross section $x = 0$ of the neck we can accept the elasticity modulus of one fibers that is

$$E_f(0) = E_{fibre} \tag{3.144}$$

Then allowing for (3.144), formula (3.142) takes the following final form:

$$E_f(x) = E_f(\ell_0)[1 + (\frac{E_{fibre}}{E_f(\ell_0)} - 1)(1 - \frac{x}{\ell_0})] \text{ for } 0 \le x \le \ell_0 \tag{3.145}$$

Thus, the dependence of the elasticity modulus $E_f(x)$ on longitudinal coordinate x in the interval $0 \le x \le \ell_0$ will characterize the heterogeneity of cross section points on the neck's length that will change in the interval:

$$E_f(\ell_0) \le E_f(x) \le E_{fibre} \text{ for } 0 \le x \le \ell_0 \tag{3.146}$$

In formula (3.141), the coefficient $\lambda_v(x)$ characterizes the lateral contraction type effect of a polymerized bar from fibers on its lateral surface in the direction of the normal $\vec{V}$ to the contour of curved neck with regard to influence of all kinds of deformations, of the form:

$$\lambda_v(x) = -\frac{\varepsilon_v}{\varepsilon_x} = -\frac{\vec{\varepsilon}_v \vec{V}}{\vec{\varepsilon}_x \vec{i}} \tag{3.147}$$

For establishing its mathematical dependence we behave as follows. Let the numerical values of this parameter in the section $x = 0$ and $x = \ell_0$ of the neck of a polymerized bar be given that is the coefficients $\lambda_v(0)$ and $\lambda_v(\ell_0)$ be given. Here, $\lambda_v(\ell_0)$ is the value of the coefficient $\lambda_v(x)$ at the cross section $x = \ell_0$ of the neck, that is, at the end of the neck. The value of the coefficient is given in the narrow part of the neck. By the smallness of the influence of lateral deformation of a binding material in the narrow part of the neck $\lambda_v(0)$ of the bar on lateral deformation of a polymerized bar we can assume that in the narrow part of the neck $x = 0$ only the fibers directed strongly in longitudinal direction work. Therefore, in place of the coefficient $\lambda_v(0)$ at the narrow part of the neck we can take the Poisson ratio of one fiber $\lambda_v(fibre)$. Under these conditions, the dependence of the coefficient $\lambda_v(x)$ at any cross section x of the neck on longitudinal coordinate x may be represented in the following linear form:

$$\lambda_v(x) = \lambda_v(\text{fibre}) + \frac{x}{\ell_0}[\lambda_v(\ell_0) - \lambda_v(\text{fibre})] \qquad (3.148)$$

For $\lambda_v(\text{fibre}) = 0$ we get the following simplified dependence:

$$\lambda_v(x) = \lambda_v(\ell_0)\frac{x}{\ell_0} \qquad (3.149)$$

Note that the value of the coefficient at the neck's end $\lambda_v(\ell_0)$ will be the value of the coefficient $\lambda_v(x)$ out of the neck, whose mathematical expression is represented in (Aliyev, 1987, 2011, 2012) and in conformity to our problem will be of the form:

$$\lambda_v(\ell_0) = \frac{1}{v_b}[1 - \frac{1}{n}\frac{E_{averag.} - E_b(1 - nk_F)}{E_0 k_F k_\sigma}] \qquad (3.150)$$

Here $E_{averag.}$ and E_b are the elasticity module of the sample and binding material, v_b is the Poisson ratio of the binding material, n is the amount of bundles in the sample, $k_F = F_f / F$ is the ratio of the cross section area of a polymerized bundle to the cross section area of the sample, $k_\sigma = \frac{\sigma_f^{in\,medium}}{\sigma_f^{free}}$ is the coefficient characterizing the ratio of the ultimate strength of a fibrous bundle in the medium to the ultimate strength of the bundle tested in air, $F_b = F - nF_f$ is the cross-section area of the binding material of the sample.

Thus, the dependence of the coefficient $\lambda_v(x)$ on longitudinal coordinate x in the interval $0 \le x \le \ell_0$ will change in the interval:

$$0 \leq \lambda_v(x) \leq \lambda_v(\ell_0) \text{ for } 0 \leq x \leq \ell_0 \tag{3.151}$$

In formula (3.151) the coefficient $\lambda_\tau(x)$ characterizes a shear effect arising between the reinforcing fibrous structure of the bar and a binding material in the neck zone. Therefore, we can represent the mechanical characteristics of the coefficient $\lambda_\tau(x)$ at the lateral surface points of the neck as the ratio of the shear modulus of the binding material G_b to the longitudinal elasticity modulus $E_f(x)$ of a polymerized bar, of the form:

$$\lambda_\tau(x) = \frac{G_b}{E_f(x)} = \frac{G_b}{E_f(\ell_0)[1 + (\frac{E_{fibre}}{E_f(\ell_0)} - 1)(1 - \frac{x}{\ell_0})]} \tag{3.152}$$

Thus, the dependence of the coefficient $\lambda_\tau(x)$ on longitudinal coordinate x in the interval $0 \leq x \leq \ell_0$ will change in the interval:

$$\frac{G_b}{E_{fibre}} \leq \lambda_\tau(x) \leq \frac{G_b}{E_f(\ell_0)} \text{ for } 0 \leq x \leq \ell_0 \tag{3.153}$$

Thus, having experimentally defined mechanical characteristics of a polymerized bar $E_f(x)$, $\lambda_v(x)$ and $\lambda_\tau(x)$ at the cross sections of the neck by formulae (3.142) (3.149) and (3.152), we can suggest the dependence of the longitudinal stress vector $\vec{\sigma}_f(x)$ in a polymerized bar on longitudinal deformation vector $\vec{\varepsilon}_x = \varepsilon_x \vec{i}$ and deformation vectors $\vec{\varepsilon}_v$ and $\vec{\varepsilon}_\tau$ acting on lateral surface of the bundle at the neck's cross sections, in the form of formula (3.141).

Example: On an example of a polymerized bar made on the base of filamentary structure from a glass fiber and a rubber, we give the method for experimental definition of mechanical characteristics of a polymerized glass bar $E_f(x)$, $\lambda_v(x)$ and $\lambda_\tau(x)$ at the neck zone and suggest specific mechanical models of deformation of glass rubber bars at the neck zone.

We'll numerically show the degree of influence of stress state of a binding medium (rubber) on mechanical characteristics of a bundle shaped structure at the neck zone. To this end we have experimentally investigated the glass braids of the brands BS6-34 × 1 × 3 × n. Here the glass filament consists of 400 elementary glass fibers of the Tex BS6-34 ×1. The influence of the stress-state form of a binding material (rubber) on mechanical properties of a glass bundle of the given structure in which they are reinforced, is studied.

For $E_{rubber} = 80 \, {}^{kgf}\!/_{cm^2}$, $\nu_{rubber} = 0,47$

TABLE 3.2 The results of tests

Brand of bundle from glass fiber	**BS6 × 34 ×1 ×3 × 5**	**BS6 × 34 ×1 ×3 × 10**	**BS6 × 34 ×1 ×3 × 15**
Cross section area of a bundle $F_f(cm^2)$	$237\cdot10^{-5}$	$465\cdot10^{-5}$	$685\cdot10^{-5}$
Elasticity modulus of a bundle $E_0\left(kgf/cm^2\right.$	$0.524\cdot10^{6}$	$0.442\cdot10^{6}$	$0.427\cdot10^{6}$
Ultimate strength of a glass bundle without influence of binding medium $\sigma^0_{6H}\left({}^{kgf}\!/_{cm^2}\right)$	6,751	6,451	6,258
Cross section area of the sample $F\left(cm^2\right)$	1.33	1.33	1.33
$k_F = \frac{F_f}{F}$	$178\cdot10^{-5}$	$350\cdot10^{-5}$	$516\cdot10^{-5}$

TABLE 3.2 *(Continued)*

Ultimate strength of a glass bundle in rubber $\sigma_{ef}(\text{kgf}/\text{cm}^2)$	8,439	8,172	8,105
$k_\sigma = \dfrac{\sigma_{ef}}{\sigma_{ef}^0}$	1.2500	1.2668	1.2931
$E_{cp}\left(\text{kgf}/\text{cm}^2\right)$	872.38	1501.93	2234.92
Elasticity modulus of a bundle in rubber $E_f(\ell_0)\left(\text{kgf}/\text{cm}^2\right)$	$0.655 \cdot 10^6$	$0.561 \cdot 10^6$	$0.551 \cdot 10^6$
Averaged coefficient of lateral compression of a fiber bundle $\lambda_v(\ell_0)$	0.684	0.585	0.512

The thick walled stripes with cross section area 1.33cm^2 along which the glass braids in the amount of 1, 3, 5 were reinforced, were tested. The elasticity modulus $E_f(\ell_0)$ averaged coefficient of lateral compression of a glass fire in the rubber $\lambda_v(\ell_0)$ and ultimate strength of the reinforced glass braid with regard to stress state of the binder (rubber) $\sigma_f^{in\,medium}$ out of neck were determined on the base of experimental data. The results of tests are given in Table 3.2:

$$\sigma_f^5(\ell_0) = 0.655 \cdot 10^6(\varepsilon_x + 0.684\varepsilon_\perp)\text{, for brand } BS6 \times 34 \times 1 \times 3 \times 5$$

$$\sigma_f^{10}(\ell_0) = 0.561 \cdot 10^6(\varepsilon_x + 0.585\varepsilon_\perp)\text{, for brand } BS6 \times 34 \times 1 \times 3 \times 10$$

$$\sigma_f^{15}(\ell_0) = 0.551 \cdot 10^6(\varepsilon_x + 0.512\varepsilon_\perp)\text{, for brand } BS6\times34\times1\times3\times15 \quad (3.154)$$

$$\text{For } E_{fibre} = 0.8 \cdot 10^6 \frac{\text{kgf}}{\text{cm}^2}$$

TABLE 3.3 Mathematical dependence of mechanical characteristics of glass braids polymerized in rubber at the neck zone

$\sigma_f^m(x)$	$\lambda_v^m(x)$	$\lambda_\tau^m(x)$
$E_f^5(x) = 0.655 \cdot 10^6[1 + 0.2213(1 - \frac{x}{\ell_0})]$	$\lambda_v^5(x) = 0.684\frac{x}{\ell_0}$	$\lambda_\tau^5(x) = \frac{41.542}{1 + 0.2213(1 - \frac{x}{\ell_0})} 10^{-6}$
$E_f^{10}(x) = 0.561 \cdot 10^6[1 + 0.4260(1 - \frac{x}{\ell_0})]$	$\lambda_v^{10}(x) = 0.585\frac{x}{\ell_0}$	$\lambda_\tau^{10}(x) = \frac{48.503}{1 + 0.4260(1 - \frac{x}{\ell_0})} 10^{-6}$
$E_f^{15}(x) = 0.551 \cdot 10^6[1 + 0.4519(1 - \frac{x}{\ell_0})]$	$\lambda_v^5(x) = 0.512\frac{x}{\ell_0}$	$\lambda_\tau^{15}(x) = \frac{49.383}{1 + 0.4519(1 - \frac{x}{\ell_0})} 10^{-6}$

From formulae (3.142), (3.149), and (3.152), with regard to experimental data of Table 3.1 and formula (3.154), we establish the following mathematical dependence of mechanical characteristics of glass braids polymerized in rubber at the neck zone $E_f(x)$, $\lambda_v(x)$ and $\lambda_\tau(x)$ on longitudinal coordinate x, that are represented in Table 3.3.

The peculiarity of these mechanical characteristics is the following:

elasticity modulus of a glass fiber polymerized in a rubber $E_f(x)$ at the narrow part of the neck is independent of quantity content in it elementary fibers and equals:

$$E_f^m(x)\Big|_{x=0} = 0.8 \cdot 10^6 \,\text{kgf}/\text{cm}^2 \quad (3.155)$$

at the neck zone, the longitudinal elasticity modulus $E_f(x)$ depending on x changes in the following form:

$$0.7999\cdot 10^6\,\frac{kgf}{cm^2}\geq E_f^5(x)\geq 0.655\cdot 10^6\,\frac{\text{kgf}}{\text{cm}^2}$$

$$0.8\cdot 10^6\,\frac{kgf}{cm^2}\geq E_f^{10}(x)\geq 0.561\cdot 10^6\,\frac{\text{kgf}}{\text{cm}^2},\ \text{for } 0\leq x\leq \ell_0$$

$$0.8\cdot 10^6\,\frac{kgf}{cm^2}\geq E_f^{15}(x)\geq 0.551\cdot 10^6\,\frac{\text{kgf}}{\text{cm}^2} \tag{3.156}$$

Hence, it is seen that changeability of elasticity modulus of a glass bundle polymerized in a rubber, is essential at the neck zone depending on x. Comparing the values of elasticity module of a polymerized bundle at the narrowest place with the elasticity modulus at the neck's end (out of neck) we get: for a 5 fold glass bundle by 22.12% 10 fold by 42.6% and 15 fold by 45.2% greater than the appropriate elasticity module at the neck end.

averaged coefficient of lateral contraction of a glass fiber polymerized in a rubber. $\lambda_v(\ell_0)$ at the neck's end essentially depends on quantity content of elementary fibers in it that equal: $\lambda_v^5(\ell_0)=0.684$, $\lambda_v^{10}(\ell_0)=0.585$, $\lambda_v^{15}(\ell_0)=0.512$.

at the neck zone, the averaged coefficient of lateral contraction of a glass fiber polymerized in a rubber $\lambda_v(x)$ depending on x $(0\leq x\leq \ell_0)$ changes as follows:

$$0\leq \lambda_v^5(x)\leq 0{,}684,\ 0\leq \lambda_v^{10}(x)\leq 0{,}584,\ 0\leq \lambda_v^{15}(x)\leq 0{,}512 \tag{3.157}$$

the Shael coefficient $\lambda_\tau(x)$ between the reinforcing fibrous structure of the bar and binding rubber at the neck zone have the following peculiarities.

The Shael coefficient $\lambda_\tau(x)$ at the narrow part of the neck is independent of quality content of elementary glass fibers in it and equals:

$$\lambda_\tau^{m}(x)\Big|_{x=0} = 34 \cdot 10^{-6} \tag{3.158}$$

and at the neck's end the coefficient $\lambda_\tau(x)$ has the following numerical values:

$$\lambda_\tau^{5}(\ell_0) = 41.542 \cdot 10^{-6}, \quad \lambda_\tau^{10}(\ell_0) = 48.503 \cdot 10^{-6},$$

$$\lambda_\tau^{5}(\ell_0) = 49.383 \cdot 10^{-6} \tag{3.159}$$

The coefticient $\lambda_\tau(x)$ at the cross sertions of the neck zone depending on x $(0 \le x \le \ell_0)$ changes in the following form:

$$\lambda_\tau^{5}(x) = \frac{41.542}{1 + 0.2213(1 - \frac{x}{\ell_0})} \cdot 10^{-6},$$

$$\lambda_\tau^{10}(x) = \frac{48.503}{1 + 0.4260(1 - \frac{x}{\ell_0})} \cdot 10^{-6}$$

$$\lambda_\tau^{15}(x) = \frac{49.383}{1 + 0.4519(1 - \frac{x}{\ell_0})} \cdot 10^{-6} \tag{3.160}$$

On the basis of these experimental data we suggest the following models of mechanical deformation of a glass bundles (filament) polymerized in a rubber at the neck zone depending on x $(0 \le x \le \ell_0)$:
for the brand $BS6 \times 34 \times 1 \times 3 \times 5$:

$$\vec{\sigma}_f^5(x) = 0.655 \cdot 10^6 \cdot [1 + 0.2213(1 - \frac{x}{\ell_0})]\{\vec{\varepsilon}_x + 0.684\frac{x}{\ell_0} \cdot \vec{\varepsilon}_v +$$

$$+ \frac{41.542 \cdot 10^{-6}}{1 + 0.2213(1 - \frac{x}{\ell_0})} \cdot \vec{\varepsilon}_\tau\}$$

for the brand $BS6 \times 34 \times 1 \times 3 \times 10$:

$$\vec{\sigma}_f^{10}(x) = 0.561 \cdot 10^6 \cdot [1 + 0.4260(1 - \frac{x}{\ell_0})]\{\vec{\varepsilon}_x + 0.585\frac{x}{\ell_0} \cdot \vec{\varepsilon}_v +$$

$$+ \frac{48.503 \cdot 10^{-6}}{1 + 0.4260(1 - \frac{x}{\ell_0})} \cdot \vec{\varepsilon}_\tau\}$$

for the brand $BS6 \times 34 \times 1 \times 3 \times 15$:

$$\vec{\sigma}_f^{15}(x) = 0.551 \cdot 10^6 \cdot [1 + 0.4519(1 - \frac{x}{\ell_0})]\{\vec{\varepsilon}_x +$$

$$+0.512\frac{x}{\ell_0} \cdot \vec{\varepsilon}_v + \frac{49.383 \cdot 10^{-6}}{1 + 0.4519(1 - \frac{x}{\ell_0})} \cdot \vec{\varepsilon}_\tau\} \quad (3.160)$$

3.5 EXPERIMENTAL—THEORETICAL METHOD FOR DEFINING PHYSICO–MECHANICAL PROPERTIES OF A POLYMERIZED BUNDLE FROM FIBERS WITH REGARD TO CHANGE OF PHYSICO–CHEMICAL PROPERTIES OF A BINDING MATERIAL

In a composite material, one of the important problems is to provide continuity the material and adhesive strength of the construction. Influence of physicochemical change of the composite's elements arising under the action of corrosive liquid and gassy media on the continuity problem of the composites material and constrictions adhesive property is a principal moment in this problem. In references there are no fundamental investigations of such kind. Below we suggest an experimental–theoretical method of calculation of influence of physico–mechanical properties of a binding material on physico–mechanical properties of a polymerized bundles from fibers (filaments, fabrics) an construction of appropriate mechanical deformation models (Aliyev, 2012, 2012).

We suggest a model of a linear mechanical deformation of a polymerized bundle-shaped body (filament, braid, fabric) with regard to change of physico–chemical properties of a corrosive liquid or grassy medium in the binding material, in the form:

$$T_f = E_f(\lambda)F_f(\lambda)[\varepsilon_1 + v'_\perp(\lambda)\varepsilon_\perp^b + \frac{v''_\perp(\lambda)}{E_f(\lambda)}\sigma_\perp^b - \alpha_f(\lambda)\cdot\tilde{\lambda}] \quad (3.161)$$

where $E_f(\lambda)$, $v'_\perp(\lambda)$, $v''_\perp(\lambda)$ are elastic constants of a polymerized bundle from fibers in a polymer medium with regard to swelling effects of a binding material that arises from the action of corrosive liquid or grassy medium. Hence, it follows that for each parameter λ there will correspond it own values of elastic constants of a polymerized filament from fibers. Below we along will give an experimental method for their definition; ε_1 is deformation of a bundle the bundle; $\varepsilon_\perp^b = \frac{1}{2}(\varepsilon_2^b + \varepsilon_3^b)$, $\sigma_\perp^b = \frac{1}{2}(\sigma_2^b + \sigma_3^b)$ are deformation and stress lateral to the bundle and expressed by the principal quantities, $F_f(\lambda)$ in a cross section area of a bundle from fibers; λ is a swelling parameter of a binding material; $\alpha(\lambda)\cdot\lambda$ is deformation arising at the expense of swelling effect of a binding material. In particular, when there are no deformations $\varepsilon_\ell = \varepsilon_\perp = 0$ in the sample will arise initial stress from the physico–chemical change of the material in the form:

$$T_f = -E_f(\lambda)F_f(\lambda)\alpha_f(\lambda)\cdot\tilde{\lambda} \quad (3.162)$$

In other words, under the action of only swelling effect, in the polymerized bundle from fibers there will arise tensile force of quantity (3.162). For $\lambda = 0$, the suggested deformation model will coincide with a polymerized bundle model regardless of swelling effect suggested in (Aliyev, 1984, 1987).

In this section we'll consider a thin walled shell reinforced on the middle of thickness by compacted filaments sloping at angle β to the longitudinal axis (Figure 3.1). The bonds of principal stresses on the bonder σ_1^b, σ_2^b, σ_3^b with coordinate stresses for the case of a Hein walled shell $\sigma_3^b = 0$ will be expressed by the known relations (Ilyushin, 1948; Ilyushin and Lenskiy, 1959):

$$\begin{cases} \sigma_1^b = \frac{1}{2}(\sigma_{xx}^b + \sigma_{yy}^b) + \frac{1}{2}(\sigma_{xx}^b - \sigma_{yy}^b)\cos 2\beta + \sigma_{xy}^b \sin 2\beta \\ \sigma_2^b = \frac{1}{2}(\sigma_{xx}^b + \sigma_{yy}^b) - \frac{1}{2}(\sigma_{xx}^b - \sigma_{yy}^b)\cos 2\beta - \sigma_{xy}^b \sin 2\beta \\ \sigma_3^b = \frac{1}{2}(\sigma_{xx}^b - \sigma_{yy}^b)\sin 2\beta + \sigma_{xy}^b \cos 2\beta \end{cases} \tag{3.163}$$

Then the lateral stress in the binder on the boundary of contact with the reinforcing layer $\sigma_\perp^b = \frac{1}{2}(\sigma_2^b + \sigma_3^b)$ with regard to $\sigma_3^b \equiv 0$ and (3.163) will be expressed by the coordinate stresses in the form:

$$\sigma_\perp^b = \frac{1}{4}(\sigma_{xx}^b + \sigma_{yy}^b) - \frac{1}{4}(\sigma_{xx}^b - \sigma_{yy}^b)\cos 2\beta - \frac{1}{2}\sigma_{xy}^b \sin 2\beta \tag{3.164}$$

The Hooke's generalized law for a binding material with regard to change effect of its physico–chemical property has the form (1.91), that is:

$$\begin{cases} \sigma_{ij}(\varepsilon_{ij},\lambda) = a_0^b\psi_b(\lambda)\cdot\theta\cdot\delta_{ij} + 2G_0^b\phi_b(\lambda)\varepsilon_{ij}^b - \eta_0^b\eta_b(\lambda)\cdot\alpha_b\cdot\tilde{\lambda}\cdot\delta_{ij} & (3.165) \\ \sigma(\varepsilon,\lambda) = 3a_0^b\psi_b(\lambda)[1 + \frac{2G_0^b}{3a_0^b}\frac{\phi_b(\lambda)}{\psi_b(\lambda)}](\theta - 3\tilde{\alpha}_b\tilde{\lambda}_b) & (3.166) \end{cases}$$

where the mechanical properties a_0^b, G_0^b, η_0, $\tilde{\alpha}_b$ and the properties dependent on physico–chemical changes $\varphi_b(\lambda)$, $\psi_b(\lambda)$, $\eta_b(\lambda)$ have the form (1.81) and (1.86). In principal stresses and deformations they will be of the form:

$$\begin{cases} \sigma_1 = (a_0^b\psi_b + 2G_0^b\phi_b)\varepsilon_1 + a_0^b\psi_b\varepsilon_2 + a_0^b\psi_b\varepsilon_3 - \eta_0^b\eta_b(\lambda)\alpha_b\tilde{\lambda} \\ \sigma_2 = a_0^b\psi_b\varepsilon_1 + (a_0^b\psi_b + 2G_0^b\phi_b)\varepsilon_2 + a_0^b\psi_b\varepsilon_3 - \eta_0^b\eta_b(\lambda)\alpha_b\tilde{\lambda} \\ \sigma_3 = a_0^b\psi_b\varepsilon_1 + a_0^b\psi_b\varepsilon_2 + (a_0^b\psi_b + 2G_0^b\phi_b)\varepsilon_3 - \eta_0^b\eta_b(\lambda)\alpha_b\tilde{\lambda} \end{cases} \quad (3.167)$$

Solving system (3.167), we find the dependence of principal deformations by the principal stresses in the form:

$$\varepsilon_1^b = \frac{1+\frac{G_0^b}{a_0^b}\frac{\phi_b(\lambda)}{\psi_b(\lambda)}}{3G_0^b\phi_b(\lambda)[1+\frac{2}{3}\frac{G_0^b}{a_0^b}\frac{\phi_b(\lambda)}{\psi_b(\lambda)}]}[\sigma_1^b - \frac{1}{2\left(1+\frac{G_0^b}{a_0^b}\frac{\phi_b(\lambda)}{\psi_b(\lambda)}\right)}(\sigma_2^b+\sigma_3^b)]+$$

$$+\frac{1}{3a_0^b\psi_b(\lambda)[1+\frac{2}{3}\frac{G_0^b}{a_0^b}\frac{\phi_b(\lambda)}{\psi_b(\lambda)}]}\cdot\eta_0^b\eta^b(\lambda)\cdot\alpha_b\tilde{\lambda}_b$$

$$\varepsilon_2^b = \frac{1+\frac{G_0^b}{a_0^c}\frac{\phi_b(\lambda)}{\psi_b(\lambda)}}{3G_0^b\phi_b(\lambda)[1+\frac{2}{3}\frac{G_0^b}{a_0^b}\frac{\phi_b(\lambda)}{\psi_b(\lambda)}]}[\sigma_2^b - \frac{1}{2\left(1+\frac{G_0^b}{a_0^b}\frac{\phi_b(\lambda)}{\psi_b(\lambda)}\right)}(\sigma_1^b+\sigma_3^b)]+$$

$$+\frac{1}{3a_0^b\psi_b(\lambda)[1+\frac{2}{3}\frac{G_0^b}{a_0^b}\frac{\phi_b(\lambda)}{\psi_b(\lambda)}]}\cdot\eta_0^b\eta^b(\lambda)\cdot\alpha_b\tilde{\lambda}_b$$

$$\varepsilon_3^b=\frac{1+\frac{G_0^b}{a_0^b}\frac{\phi_b(\lambda)}{\psi_b(\lambda)}}{3G_0^b\phi_b(\lambda)[1+\frac{2}{3}\frac{G_0^b}{a_0^b}\frac{\phi_b(\lambda)}{\psi_b(\lambda)}]}[\sigma_3^b$$

$$-\frac{1}{2\left(1+\frac{G_0^b}{a_0^b}\frac{\phi_b(\lambda)}{\psi_b(\lambda)}\right)}(\sigma_1^b+\sigma_2^b)]+$$

$$+\frac{1}{3a_0^b\phi_b(\lambda)[1+\frac{2}{3}\frac{G_0^b}{a_0^b}\frac{\phi_b(\lambda)}{\psi_b(\lambda)}]}\cdot\eta_0^b\eta^b(\lambda)\cdot\alpha_b\tilde{\lambda}_b \tag{3.168}$$

Here, a_0^b G_0^b are the Lame coefficients of a binding material for $\lambda=0$, that is regardless of swelling effect; $\varphi_b(\lambda)$, $\psi_b(\lambda)$, $\eta^b(\lambda)$ are correction functions.

For the case of thin-walled shell $\sigma_3^b\equiv 0$, from the third relation of (3.168) we have:

$$\varepsilon_r=\varepsilon_3^b=-\frac{1}{3G_0^b\phi_b(\lambda)[1+\frac{2}{3}\frac{G_0^b}{a_0^b}\frac{\phi_b(\lambda)}{\psi_b(\lambda)}]}[\frac{1}{2}(\sigma_1^b+\sigma_2^b)$$

$$-\frac{G_0^b}{a_0^b}\frac{\phi_b(\lambda)}{\psi_b(\lambda)}\cdot\eta_0^b\eta^b(\lambda)\cdot\alpha_b\tilde{\lambda}_b] \tag{3.169}$$

Substituting (3.163) in (3.169), define the deformation $\varepsilon_r=\varepsilon_3^b$ by the coordinate stresses:

$$\varepsilon_r = \varepsilon_3^b = -\frac{1}{3G_0^b\phi_b(\lambda)[1+\frac{2}{3}\frac{G_0^b}{a_0^b}\frac{\phi_b(\lambda}{\psi_b(\lambda)}]}[\frac{1}{2}(\sigma_{xx}^b+\sigma_{yy}^b)-$$

$$\frac{G_0^b}{a_0^b}\frac{\phi_b(\lambda)}{\psi_b(\lambda b}\cdot\eta_0^b\eta^b(\lambda)\cdot\alpha_b\tilde{\lambda}_b] \qquad (3.170)$$

For finding lateral deformation $\varepsilon_\perp$ and lateral stress $\sigma_\perp$ by the coordinate deformation for the case of a shell, with regard to swelling, we use the first two equations of (3.168) and find:

$$\sigma_x^b = 6G_0^b\phi_b(\lambda)\frac{k_2(\phi,\psi)\cdot k_3(\phi,\psi)}{1-k_2^2(\phi,\psi)}[\varepsilon_x^b + k_2(\phi,\psi)\varepsilon_y^b]$$

$$-2\frac{G_0^b}{a_0^b}\frac{\phi_b(\lambda)}{\psi_b(\lambda)}\cdot\eta_0^b\eta_b(\lambda)\cdot\alpha_b\tilde{\lambda}_b\frac{k_2(\phi,\psi)}{1-k_2(\phi,\psi)} \qquad (3.171)$$

$$\sigma_y^b = 6G_0^b\phi_b(\lambda)\frac{k_2(\phi,\psi)\cdot k_3(\phi,\psi)}{1-k_2^2(\phi,\psi)}[\varepsilon_y^b + k_2(\phi,\psi)\varepsilon_x^b]$$

$$-2\frac{G_0^b}{a_0^b}\frac{\phi_b(\lambda)}{\psi_b(\lambda)}\cdot\eta_0^b\eta_b(\lambda)\cdot\alpha_b\tilde{\lambda}_b\frac{k_2(\phi,\psi)}{1-k_2(\phi,\psi)} \qquad (3.172)$$

Here:

$$\sigma_{xy}^b = 2G_0^b\phi_b(\lambda)\cdot\varepsilon_{xy}^b \qquad (3.173)$$

$$k_2(\phi,\psi) = -\frac{1}{2(1+\frac{G_0^b}{a_0^b}\frac{\phi_b(\lambda}{\psi_b(\lambda)})},$$

$$k_3(\phi,\psi)=1+\frac{2}{3}\frac{G_0^b}{a_0^b}\frac{\phi_b(\lambda)}{\psi_b(\lambda)} \tag{3.174}$$

Substituting (3.171 and (3.173) in (3.164), express the lateral stress in the binder $\sigma_\perp^b$ by the coordinate deformations ε_{xx}, ε_{yy}, ε_{xy} and swelling parameter λ in the form:

$$\sigma_\perp^b=\frac{3}{2}G_0^b\phi_b(\lambda)\frac{k_2^bk_3^b}{1-k_2^{b2}}[(1+k_2^b)(\varepsilon_x^b+\varepsilon_y^b)$$
$$-(1-k_2^b)(\varepsilon_x^b-\varepsilon_y^b)\cos 2\beta]-G_0^b\phi_b(\lambda)\varepsilon_{xy}^b\sin 2\beta$$
$$-\frac{G_0^b}{a_0^b}\frac{\phi_b(\lambda)}{\psi_b(\lambda)}\eta_0\eta_b(\lambda)\alpha_b\tilde{\lambda}_b\frac{k_2^b}{1-k_2^b} \tag{3.175}$$

Express the principal deformations ε_1^0, ε_2^0, ε_3^0 by the coordinate deformations ε_{xx}^0, ε_{yy}^0, ε_{xy}^0.

$$\begin{cases}\varepsilon_1^b=\frac{1}{2}(\varepsilon_{xx}^b+\varepsilon_{yy}^b)+\frac{1}{2}(\varepsilon_{xx}^b-\varepsilon_{yy}^b)\cos 2\beta+\frac{1+k_2^b}{3k_2^bk_3^b}\varepsilon_{xy}^b\sin 2\beta\\ \varepsilon_2^b=\frac{1}{2}(\varepsilon_{xx}^b+\varepsilon_{yy}^b)-\frac{1}{2}(\varepsilon_{xx}^b-\varepsilon_{yy}^b)\cos 2\beta-\frac{1+k_2^b}{3k_2^bk_3^b}\varepsilon_{xy}^b\sin 2\beta\\ \varepsilon_3^b=-\frac{k_2^b}{1-k_2^b}[\varepsilon_{xx}^b+\varepsilon_{yy}^b-\frac{1+k_2^b}{k_2^bk_3^b}\frac{\eta_0^b\eta_b(\lambda)\alpha_b\tilde{\lambda}_b}{3a_0\psi(\lambda)}\eta_0\eta_b(\lambda)\alpha_b\tilde{\lambda}_b]\end{cases} \tag{3.176}$$

Then the lateral deformation in the binder $\varepsilon_\perp^b=\frac{1}{2}(\varepsilon_2+\varepsilon_3)$ with regard to swelling effect λ of the binding material will equal:

$$2\varepsilon_\perp^b=\frac{1-3k_2^b}{2(1-k_2^b)}(\varepsilon_{xx}^b+\varepsilon_{yy}^b)-\frac{1}{2}(\varepsilon_{xx}^b-\varepsilon_{yy}^b)\cos 2\beta$$
$$-\frac{1+k_2^b}{3k_2^bk_3^b}\varepsilon_{xy}^b\sin 2\beta+\frac{1+k_2^b}{k_3^b(1-k_2^b)}\frac{\eta_0^b\eta_b(\lambda)\alpha_b\tilde{\lambda}_b}{3a_0\psi(\lambda)} \tag{3.177}$$

Substituting (3.175)–(3.177) in (3.161) we get a deformation model of a polymerized bundle-shaped body with regard to swelling effect depending of coordinate deformations:

$$\sigma_f = \frac{T_f}{(EF)_f} = \frac{1}{2}[\left(1 + \frac{1-3k_2^b}{2(1-k_2^b)}\nu_\perp' + 3G_0^b\phi_b(\lambda)\frac{k_2^b k_3^b}{1-k_2^b}\frac{\nu_\perp''}{E_f}\right)$$

$$+\left(1 - \frac{\nu_\perp'}{2} - 3G_0^b\phi_b(\lambda)k_2^b k_3^b\frac{\nu_\perp''}{E_f}\right)\cos 2\beta]\cdot\varepsilon_x$$

$$+\frac{1}{2}[\left(1 + \frac{1-3k_2^b}{2(1-k_2^b)}\nu_\perp' + 3G_0^b\phi_b(\lambda)\frac{k_2^b k_3^b(1+k_2^b)}{1-k_2^b}\frac{\nu_\perp''}{E_f}\right)$$

$$+\left(-1 + \frac{\nu_\perp'}{2} - 3G_0^b\phi_b(\lambda)k_2^b k_3^b\frac{\nu_\perp''}{E_f}\right)\cos 2\beta]\cdot\varepsilon_y$$

$$+\frac{1+k_2^b}{3k_2^b k_3^b}\left(1 - \frac{\nu_\perp'}{2} - G_0^b\phi_b(\lambda)\frac{3k_2^b k_3^b}{1+k_2^b}\frac{\nu_\perp''}{E_f}\right)\sin 2\beta\cdot\varepsilon_{xy}^b$$

$$+\frac{1}{a_0^b\psi_b(\lambda)(1-k_2^b)}[\frac{1+k_2^b}{2k_3^b}\nu_\perp' - \frac{G_0^b\phi_b(\lambda)k_2^b}{E_f}\nu_\perp'']$$

$$\cdot\eta_0^b\eta_b(\lambda)\alpha_b\tilde{\lambda}_b - \alpha_f\lambda \qquad (3.178)$$

For $\lambda = 0$, that is, when there is no swelling effect, the deformation model of polymerized bundle (3.178) coincides with (3.18).

Special case: For the case of longitudinal reinforcement, for $\beta = 0$, where ε_y is expressed by $\varepsilon_\perp^b$ and swelling parameter λ_b by means of

(3.175), from (3.178), it is easy to get dependence of deformation of polymerized bundle (filament) $\sigma = \sigma(\varepsilon_1, \varepsilon_\perp)$ on longitudinal $\varepsilon = \varepsilon_{xx}^b$ and lateral stresses $\sigma_\perp^b$ in the form:

$$\sigma_f = \frac{1}{2}[2 + 3G_0^b \phi_b(\lambda) \frac{k_2^b(\phi,\psi) k_3^b(\phi,\psi)}{1 - 2k_2^b(\phi,\psi)} \frac{v''_\perp}{E_f}] \cdot \varepsilon_x$$

$$+[v'_\perp + 6G_0^b \phi_b \frac{k_2^{b2}(\phi,\psi) k_3^b(\phi,\psi)}{1 - 2k_2^b(\phi,\psi)} \frac{v''_\perp}{E_f}] \cdot \varepsilon_\perp$$

$$+[\frac{1 + k_2^b(\phi,\psi)}{3k_3^b(\phi,\psi)(1 - k_2^b(\phi,\psi))} v'_\perp$$

$$-G_0^b \phi_b(\lambda) \frac{k_2^b(\phi,\psi)(1 - k_2^b(\phi,\psi) + k_2^{b2}(\phi,\psi))}{(1 - k_2^b(\phi,\psi))(1 - 2k_2^b(\phi,\psi))}] \cdot$$

$$\cdot \frac{\eta_0^b \eta_b(\lambda) \alpha_b \tilde{\lambda}_b}{a_0^b \psi_b(\lambda)} - \alpha_f \lambda \qquad (3.179)$$

or

$$\sigma_f = [1 - k_2^b(\phi,\psi) \cdot v'_\perp + +3G_0^b \frac{k_2^{b2}(\phi,\psi) k_3^b(\phi,\psi)(1 - 2k_2^b(\phi,\psi)}{2(1 - k_2^b(\phi,\psi))} \frac{v''_\perp}{E_f}] \cdot \varepsilon_x$$

$$+\frac{1 + k_2^b(\phi,\psi)}{6G_0^b \phi_c k_2^b(\phi,\psi) k_3^b(\phi,\psi)} [(1 - 2k_2^b(\phi,\psi)) v'_\perp$$

$$+6G_0^b\phi_b(\lambda)k_2^{b2}(\phi,\psi)k_3^b(\phi,\psi)\frac{v''_\perp}{E_f}]\cdot\sigma_\perp^b$$

$$+[\frac{1+k_2^b(\phi,\psi)}{k_3^b(\phi,\psi)}v'_\perp$$

$$-G_0^b\phi_b(\lambda)k_2^b(\phi,\psi)\frac{3k_2^b(1-k_2^b(\phi,\psi))k_2^b(\phi,\psi)-1+k_2^b(\phi,\psi)}{1-k_2^{b2}(\phi,\psi)}]$$

$$\cdot\frac{\eta_0^b\eta_b(\lambda)\alpha_b\tilde{\lambda}_b}{a_0^b\psi_b(\lambda)}-\alpha_f\lambda \tag{3.180}$$

Here, $v'_\perp$, $v''_\perp$, $E_f(\lambda)$, α_f are the constants of a polymerized bundle that should be determined in a polymer medium with regard to swelling effect of a binding material.

3.5.1 A WAY FOR EXPERIMENTAL DEFINITION OF ELASTIC CONSTANTS OF A POLYMERIZED FILAMENT $E_f(\lambda)$, $v'_\perp$, $v''_\perp$, $E_f(\lambda)$, α_f WITH REGARD TO SWELLING EFFECT OF A BINDING MATERIAL.

For defining the constants of a polymerized bundle-shaped body (filament, fabric, and braid) $E_f(\lambda)$, $v'_\perp$, $v''_\perp$, α_f depending on joint action of arising interaction forces and swelling parameter, it is necessary to know the mechanical properties of a binding material and composite sample for different values of swelling parameter λ_f. Then we find the property of a polymerized bundle in the binder by the subtraction that is

$$P_\zeta = P'_\zeta - P''_\zeta, \quad P_\eta = P'_\eta - P''_\eta, \quad P_\alpha = P'_\alpha - P''_\alpha$$

Below we suggest a variant for experimental definition of elastic constants and influence functions from the swelling parameter on thin-walled tubes with one layer force layer consisting of filaments directed along the generator $\beta = 0$. In this concretion, let the following dependences be experimentally given for a binder and a composite sample of the form:

$$\lambda_n = \frac{Q_0(t) - Q_b(t)}{Q_b(0)} \sim t, \quad \varepsilon_\ell^b(\lambda) = \frac{\Delta\ell_b(\lambda)}{\ell_0^b} \sim t,$$

$$\varepsilon_V^b(\lambda) = \frac{\Delta V_b(\lambda)}{V_0} \sim t,$$

$$\varepsilon_\ell^b(\lambda) = \alpha_b\lambda, \quad \varepsilon_V^b(\lambda) = 3\alpha_b\lambda$$

$$E_b(\lambda) = E_0^b \cdot b_b^{\tilde{\lambda}_b} = E_0^b \left(\frac{E_{min}^b}{E_0^b} \right)^{\lambda / \lambda_{max}^b} = \varepsilon_0^b \phi_b(\lambda) \qquad (3.180)$$

for a composite material:

$$\lambda_k = \frac{Q_k(t) - Q_k(t)}{Q_k(0)} \sim t, \quad \varepsilon_\ell^k(\lambda) = \frac{\Delta\ell_k(\lambda)}{\ell_0^k} \sim t,$$

$$\varepsilon_F^k(\lambda) = \frac{\Delta F_b(\lambda)}{F_0^k} \sim t, \quad \varepsilon_V^k(\lambda) = \frac{\Delta V_k(\lambda)}{V_0} \sim t,$$

$$\varepsilon_\ell^k(\lambda) = \alpha_b\lambda, \quad \varepsilon_V^k(\lambda) = 3\alpha_k\lambda,$$

$$E_k(\lambda) = E_0^k \cdot b_k^{\tilde{\lambda}_k} = E_0^k \left(\frac{E_{min}^k}{E_0^k} \right)^{\lambda / \lambda_{max}^k} = \varepsilon_0^k \phi_k(\lambda) \tag{3.182}$$

The experimental definition method of these dependences was given in the Chapter 1.

For defining mechanical constants of a polymerized filament, we have:

$$E_f(\lambda) = E_f^o \cdot \phi_f(\lambda), \ \ v'_{\perp f}(\lambda) = v'_{\perp 0} \cdot \phi'_{\perp}(\lambda),$$

$$v''_{\perp f}(\lambda) = v''_{\perp 0} \cdot \phi''_{\perp}(\lambda), \ \ \alpha_f$$

It is necessary to carry out the following experiments A thin-walled tube with a layer of reinforcing filaments directed along the tube is tested. In this case $\beta = 0$ (Figure 3.1). Here F_f is the cross section area of one filament; $F_k = F_f + F_b$ is the cross section area of the composite sample. For the case of o thin walled tube $\sigma_3^0 \equiv 0$ we have:

$$\beta = 0, \ \sigma_1^b = \sigma_x^b, \ \sigma_2^b = \sigma_y^b, \ \sigma_{xy}^b = 0, \ \sigma_r = \sigma_3 = \frac{p_a + p_b}{2},$$

$$\varepsilon_{xy}^b = 0, \ \varepsilon_y^b = \varepsilon_2^b, \ \varepsilon_x = \varepsilon_1^b$$

where are internal and external pressures, m is the number of filaments per the pipes section. In this case the total longitudinal force equals:

$$p = m(T_f + \sigma_1^b F_b) \tag{3.183}$$

$$\sigma_{\perp}^b = \frac{1}{2}(\sigma_y^b + \sigma_3^b) = \frac{1}{2}\sigma_y^b \tag{3.184}$$

$$\sigma_y^b h = R(p_a - p_b) \tag{3.185}$$

Test 1: Prime tension. In this case $p_a = p_b = 0$, $2\sigma_\perp^b = \sigma_y^b = 0$. According to Hooke's generalized law for binding material with regard to swelling effect (100.5) and thin-wallness of the tube $\sigma_\perp^N = 0$ we have:

$$\sigma_1^b = 6G_0\phi_b(\lambda)k_2(\phi,\psi)k_3(\phi,\psi)[\varepsilon_1^b -$$

$$-2\frac{G_0^b}{a_0^b}\frac{\phi_b(\lambda)}{\psi_b(\lambda)}k_2(\phi,\psi)\eta_0^b\eta_b(\lambda)\alpha_b\tilde{\lambda}_b] \tag{3.186}$$

In the case of tension only, a model for a polymerized filament will be:

$$T_f = E_f^o F_f(\lambda)\cdot\phi_f(\lambda)[\varepsilon_1 - \alpha_f\lambda] \tag{3.187}$$

Substituting (3.189) and (3.180) in (3.176), we get:

$$P/_m = E_f^o F_f(\lambda)\cdot\phi_f(\lambda)[\varepsilon_1^b - \alpha_f\lambda]$$
$$+6G_0^b F_b(\lambda)\phi_b(\lambda)k_2(\phi,\psi)k_3(\phi,\psi)[\varepsilon_1^b -$$

$$-2\frac{G_0^b}{a_0^b}\frac{\phi_b(\lambda)}{\psi_b(\lambda)}k_2(\phi,\psi)\eta_0^b\eta_b(\lambda)\alpha_b\tilde{\lambda}_b] = E_k^o F_k(\lambda)\cdot\phi_k(\lambda)\cdot\varepsilon_1 \tag{3.188}$$

For each value of the swelling parameter λ, we construct the diagram $\frac{P}{mF_k(\lambda)} = f(\varepsilon_1)$ and find the dependence $E_k(\lambda) = E_k^0\cdot\phi_k(\lambda)$, that is we define the behavior of the swelling influence function $\phi_k(\lambda)$. According to the experiment, the function $\phi_k(\lambda)$ has the form:

$$\phi_k(\lambda) = \left(\frac{E^k_{min}}{E^k_0}\right)^{\lambda/\lambda^k_{max}} = b_k^{\tilde{\lambda}_k} \tag{3.189}$$

whose definition method was stated in Chapter 1. Taking into account $\varepsilon^b_1 = \alpha_b \lambda$, from equation (3.5.28) we define the swelling function of a polymerized filament $\phi_f(\lambda)$:

$$\phi_f(\lambda) = \frac{1}{E^o_f F_f(\lambda)[1 - \frac{\alpha_f}{\alpha_b}]} \{E^o_k F_k(\lambda) \cdot \phi_k(\lambda)$$

$$-6G^b_0 F_b(\lambda)\phi_b(\lambda)k_2(\phi,\psi)k_3(\phi,\psi)[1$$

$$-2\frac{G^b_0}{a^b_0}\frac{\phi_b(\lambda)}{\psi_b(\lambda)}k_2(\phi,\psi)\eta^b_0\eta_b(\lambda)\frac{\lambda_b}{\lambda}]\} \tag{3.190}$$

Taking into account geometric characteristics of a composite, binder and filament of only swelling effect, that is, before application of mechanical loads, we have:

$$F_k(\lambda) = F^0_k(\lambda)(1+\varepsilon_x), \quad F_f(\lambda) = F^0_f(\lambda)(1+\varepsilon^0_f),$$

$$F_b(\lambda) = F^0_b(\lambda)(1+\varepsilon^0_b) \tag{3.191}$$

Here, F^0_k, F^0_f, F^0_f are the cross section areas regardless of swelling parameter; ε^0_k, ε^0_b, ε^0_f are deformations at the expense of swelling parameter.

Substituting (3.191) in (3.190) and assuming $\alpha_f << \alpha_b$, we get:

$$\phi_f(\lambda) = \frac{1}{E_f^0 k_F^0 (1+\varepsilon_F^0)} \{E_k^o \cdot \phi_k(\lambda)(1+\varepsilon_k^0)$$

$$-6G_0^b \cdot \phi_b(\lambda)[1+\varepsilon_k^0 - k_F(1+\varepsilon_f^0)]k_2(\phi,\psi)k_3(\phi,\psi)[1$$

$$-2\frac{G_0^b}{a_0^b}\frac{\phi_b(\lambda)}{\psi_b(\lambda)}k_2(\phi,\psi) - \frac{\eta_0^b \eta_b(\lambda)}{\lambda_{max}^b}]\} \quad (3.192)$$

In particular, for $\lambda = 0$ from (3.192) we define E_f^0, that coincides with $\phi_f(\lambda)\big|_{\lambda=0} = 1$. Here $k_F = F_f^0 / F_k^0$.

Test 2: Internal pressure tension.

In this case

$$p_a = p,\ p_b = 0,\ \sigma_r^b = \sigma_3^b = 0,\ 2\sigma_\perp = (\frac{R}{h} + \frac{1}{2})p = k_F p \quad (3.193)$$

The dependence of cubic deformation of a compose pipe on internal pressure p for different values of swelling parameter λ is experimentally constructed:

$$p = K_k(\lambda) \cdot \theta = K_k(\varepsilon_1 + 2\varepsilon_\perp),\ \theta = \varepsilon_1 + \varepsilon_2 + \varepsilon_3 = \varepsilon_1 + 2\varepsilon_\perp \quad (3.194)$$

Substituting (3.194) in (3.193), we establish relation between lateral stress $\sigma_\perp^b$ and deformation $\varepsilon_\perp^b$ with regard to experimental data:

$$\varepsilon_\perp^b = \frac{1}{aK_k}(\sigma_\perp^b - \frac{1}{2}\varepsilon_1^b) \quad (3.195)$$

The total longitudinal force equals:

$$P_{total}/m = T_f + \sigma_1^b F_b = E_k F_k \varepsilon_1 \tag{3.196}$$

Solving systems (3.196) and (3.27), with regard to Hooke's generalized law (1.74), define the dependence of the stress in a polymerized bundle $\sigma_f = T_f / F_f^0$ on longitudinal deformation ε_1^0 and lateral stress in the binder $\sigma_\perp^b$:

$$T_f / F_f = \frac{1}{k_F}\{E_k^o \cdot \phi_k(\lambda)(1+\varepsilon_k^0) - (1-k_F)(1+\varepsilon_b^0)[2G_0^b \cdot \phi_b(\lambda) \cdot \varepsilon_1^0$$
$$+2\frac{a_0^b}{a}\frac{\psi_b(\lambda)}{K_k} \cdot \sigma_\perp^b - \eta_0^b \eta_b(\lambda)\alpha_b \tilde{\lambda}_b]\} \tag{3.197}$$

Here, $\varepsilon_k^0 = \frac{\Delta F_k}{F_k^0}$, $\varepsilon_b^0 = \frac{\Delta F_b}{F_b^0}$ are the deformations of the composite and binder arising only at the expense of swelling effect of materials; F_k^0 and F_b^0 are the cross section areas of the sample and binder for $\lambda = 0$; $k_F = F_f^0 / F_k^0$. Comparing the coefficients for ε_1^c and $\sigma_\perp^c$ in (3.180) and (3.197), establish the dependence of the coefficient $\nu'_\perp(\lambda)$ and $\nu''(\lambda)_\perp$ on experimental data:

$$\nu'_\perp = \frac{y_1 x_{22} - y_2 x_{12}}{x_{11}x_{22} - x_{12}x_{21}}, \quad \nu''_\perp = \frac{y_2 x_{11} - y_1 x_{21}}{x_{11}x_{22} - x_{12}x_{21}} \tag{3.198}$$

where

$$x_{11} = -E_f^o \cdot \phi_f(\lambda) \cdot k_2(\phi, \psi),$$

$$x_{12} = E_f^0 \phi_f(\lambda) \cdot \frac{(1+k_2(\phi,\psi))(1-2k_2(\phi,\psi))}{6G_0^b \cdot \phi_b(\lambda) k_2(\phi,\psi) k_3(\phi,\psi)}$$

$$x_{22} = E_f^o \cdot \phi_f(\lambda) \cdot k_2(\phi,\psi) \cdot (1+k_2(\phi,\psi)),$$

$$x_{21} = \frac{(1+k_2(\phi,\psi))(1-2k_2(\phi,\psi))}{6G_0^b \cdot \phi_b(\lambda) k_2(\phi,\psi) k_3(\phi,\psi)},$$

$$y_1 = \frac{1}{k_2(\phi,\psi)}[E_k^o \cdot \phi_k(\lambda)(1+\varepsilon_k^0)$$
$$-2G_0^b \cdot \phi_b(\lambda)(1-k_F)(1+\varepsilon_b^0)] - E_f^o \cdot \phi_f(\lambda)$$

$$y_2 = -2\frac{1-k_F}{k_F}(1+\varepsilon_b^0)\frac{a_0^b \cdot \psi_b(\lambda)}{aK_k}$$

Thus, (3.197) is the model of a polymerized bundle from fibers with regard to change of physico–chemical properties of a binder.

3.6 STRESS–STRAIN STATE OF A SANDWICH-REINFORCED PIPE UNDER THE ACTION OF CORROSIVE LIQUID MEDIUM AND THE SYSTEM OF EXTERNAL LOADS

Let's investigate a strength problem of a sufficiently long thick-walled cylindrical pipe consisting of N concentrated layers of compacted symmetric bundles (braids, filaments) consisting of elementary fibers coiled at the angles β_K in the range $0 \le \beta_k \le 90^0$ to the generator, and $(N-1)$ concentrated binding layers of thicknesses $\Delta r_n = r_{n+1} - r_n$ (Figure 3.6) situated between force structures (Aliyev, 1987, 2012).

Let the given pipe be under the action of axial tensile force P, internal P_a and external P_b pressures transmitted by corrosive liquid.

Under the action of corrosive liquid, there happens its diffusion into binding layers, and this leads to change of physico–chemical and physico–mechanical properties of the binding material. In this connection, we first consider a problem on the strength of a sandwichly-reinforced pipe with regard to swelling effect of binding layers λ_n .

Find the quantity of tension of filaments (bundles, braids) and contact stresses between them and a binder, that define adhesive strength of the binder and filaments depending both on external forces and on degree of change of physico–mechanical properties of the binder. The problem is solved under the assumption of total adhesion between reinforcing and binding layers.

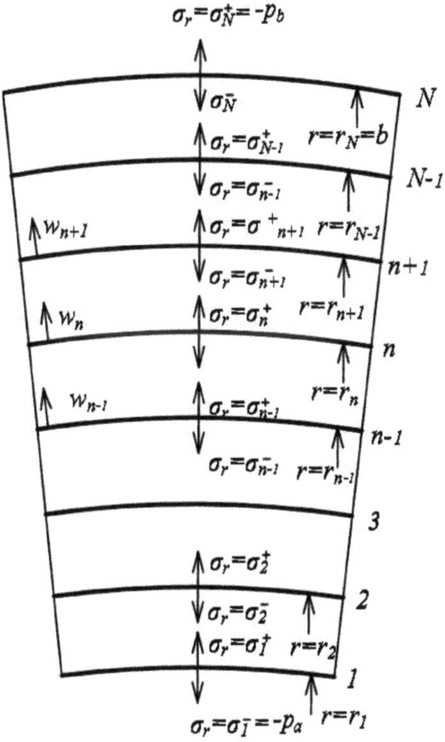

FIGURE 3.6 Multilayer-reinforced pipe with physico–chemical changeability of binder

Let's consider the n th layer of a binder ($r_n \le r \le r_{n+1}$) in the cylindrical system of coordinates and corresponding stress σ_x, σ_y, σ_r and strain ε_{xx}, ε_{yy}, ε_{rr} components. Since the pipe's structure is symmetric with respect to the axis, and its length is significantly larger than the sizes of cross section, as a basis of calculation we take the conjecture of generalized plane deformation: the lateral deformations of the pipe plane before deformation stay plane after deformation for all the values of the swelling parameter λ as well, that is the axial deformation $\varepsilon_{\tilde{o}}$ is constant along the pipe's length and they are slowly change along z functions. Then for any elastic layer we can use the relations:

$$\varepsilon_r = \frac{dw}{dr}, \ \varepsilon_y = \frac{w}{r}, \ \varepsilon_z = \varepsilon = \text{const} \tag{3.199}$$

According to (1.91) of Chapter 1, the Hooke's generalized law with regard to correction coefficients $\phi(\lambda)$, $\psi(\lambda)$, $\eta(\lambda)$ for each n-th layer ($r_n \le r \le r_{n+1}$) will have the form:

$$\sigma_{ij}^{b} = a_0 \psi_b(\lambda) \cdot \theta \cdot \delta_{ij} + 2G_0 \phi_b(\lambda) \varepsilon_{ij}^{b} - \eta_0 \eta\ (\lambda) \cdot \alpha\ \cdot \tilde{\lambda} \cdot \delta_{ij} \tag{3.200}$$

In the expanded form ($r_n \le r \le r_{n+1}$):

$$\begin{cases} \sigma_{rr} = a_0 \psi(\lambda) \cdot \theta + 2G_0 \phi(\lambda) \varepsilon_{rr} - \eta_0 \eta(\lambda) \cdot \alpha \cdot \tilde{\lambda} \\ \sigma_{yy} = a_0 \psi(\lambda) \cdot \theta + 2G_0 \phi(\lambda) \varepsilon_{yy} - \eta_0 \eta(\lambda) \cdot \alpha \cdot \tilde{\lambda} \\ \sigma_{zz} = a_0 \psi(\lambda) \cdot \theta + 2G_0 \phi(\lambda) \varepsilon_{zz} - \eta_0 \eta(\lambda) \cdot \alpha \cdot \tilde{\lambda} \end{cases} \tag{3.201}$$

Here $\theta = \varepsilon_{rr} + \varepsilon_{yy} + \varepsilon_{zz}$ is cubic deformation; $a_0 = \dfrac{E_0 v_0}{(1+v_0)(1-2v_0)}$; $2G_0 = \dfrac{E_0}{1+v_0}$ are Lame coefficients for the n-th layer of the binder regandless of swelling, that is $\lambda = 0$.

$$\phi(\lambda)=\frac{b^{\tilde{\lambda}}}{1+\dfrac{k}{\nu_0}\tilde{\lambda}},\quad \psi(\lambda)=\phi(\lambda)\frac{1+\dfrac{k}{\nu_0}\tilde{\lambda}}{1-\dfrac{2k}{1-\nu_0}\tilde{\lambda}},$$

$$\eta(\lambda)=\frac{b^{\tilde{\lambda}}\cdot\lambda_{max}}{1-\dfrac{2k}{1-2\nu_0}\tilde{\lambda}} \tag{3.202}$$

is a correction factor for a binding material dependent on the swelling parameter of the n-th binder of the pipe's layer λ; $\tilde{\lambda}=\lambda/\lambda_{max}$, $k=\nu_{max}-\nu_0$, $b=E_{min}/E_0$; α is the coefficient of linear swelling of the binding layer; E_0,ν_0 is the elasticity modulus and Poisson ratio for the binder regardless of swelling parameter $\lambda=0$.

According to Chapter 1, the dependence of volume stress $\sigma=\sigma_{ii}=\sigma_{rr}+\sigma_{yy}+\sigma_{zz}$ on total cubic deformation in the n-th layer $r_n\le r\le r_{n+1}$ (1.91), will be:

$$\sigma=3a_0\psi(\lambda)[1+\frac{2G_0}{3a_0}\frac{\phi(\lambda)}{\psi(\lambda)}](\theta-3\alpha\lambda) \tag{3.203}$$

The differential equilibrium equation of the elementary volume of the n-th layer in the radial direction ($r_n\le r\le r_{n+1}$):

$$\frac{d\sigma_{rr}}{dr}=\frac{\sigma_{yy}-\sigma_{rr}}{r} \tag{3.204}$$

Substituting (3.201) in (3.204) and taking into account (3.199) the equation for each layer of the binder is reduced to the form:

$$\frac{dw}{dr}+\frac{w}{r}=c(\lambda) \tag{3.205}$$

whose common integral will be:

$$w(r) = Ar + \frac{B}{r} \tag{3.206}$$

Then deformations (3.199) will take the form:

$$\varepsilon_{yy} = \frac{w}{r} = A + \frac{B}{r^2}, \quad \varepsilon_{rr} = \frac{\partial w}{\partial r} = A - \frac{B}{r^2}, \quad \varepsilon_z = \varepsilon = \text{const} \tag{3.207}$$

The constants A and B are defined by the boundary conditions for the n-th layer of the binder ($r_n \le r \le r_{n+1}$):

$$r = r_n, \quad w = w_n$$

$$r = r_{n+1}, \quad w = w_{n+1} \tag{3.208}$$

where w_n and w_{n+1} are the deflections of the contour points of the n-th layer. Denote by

$$\varepsilon_{yy} = \left.\frac{w}{r}\right|_{r=r_n} = \varepsilon_n, \quad \varepsilon_{yy} = \left.\frac{w}{r}\right|_{r=r_{n+1}} = \varepsilon_{n+1}$$

Satisfying conditions (3.208), determine the deformations of the points of the n-th layer by the deflections w_n and w_{n+1} on the contours:

$$\begin{pmatrix} \varepsilon_{yy} \\ \varepsilon_{rr} \end{pmatrix} = \frac{\varepsilon_{n+1} r_{n+1}^2 - \varepsilon_n r_n^2}{r_{n+1}^2 - r_n^2} \pm \frac{1}{r^2} \frac{\varepsilon_{n+1} - \varepsilon_n}{r_{n+1}^{-2} - r_n^{-2}}, \tag{3.209}$$

moreover

$$A = \frac{(wr)_{n+1} - (wr)_n}{r_{n+1}^2 - r_n^2}, \quad B = \frac{(w/r)_{n+1} - (w/r)_n}{r_{n+1}^{-2} - r_n^{-2}} \tag{3.210}$$

r is a current radius. Substituting (3.209) in (3.201), determine stresses at the points of the binder with regard to the effect of change of physico–chemical properties of the binder:

$$\frac{1}{2G_0\phi(\lambda)}\begin{pmatrix}\sigma_{yy}\\ \sigma_{rr}\end{pmatrix}=\frac{\nu_0}{1-2\nu_0}\frac{\psi(\lambda)}{\phi(\lambda)}\varepsilon_z+$$

$$+\left(1+\frac{2\nu_0}{1-2\nu_0}\frac{\psi(\lambda)}{\phi(\lambda)}\right)\frac{\varepsilon_{n+1}r_{n+1}^2-\varepsilon_n r_n^2}{r_{n+1}^2-r_n^2}\pm$$

$$\pm\frac{1}{r^2}\cdot\frac{\varepsilon_{n+1}-\varepsilon_n}{r_{n+1}^{-2}-r_n^{-2}}-\frac{\eta_0}{2G_0}\frac{\eta(\lambda)}{\phi(\lambda)}\alpha\tilde{\lambda}\ (\text{for } n=1,2,\ldots,N-1) \qquad (3.211)$$

$$\frac{1}{2G_0\phi(\lambda)}\sigma_{zz}=\left(1+\frac{\nu_0}{1-2\nu_0}\frac{\psi(\lambda)}{\phi(\lambda)}\right)\varepsilon_z+$$

$$+\frac{2\nu_0}{1-2\nu_0}\frac{\psi(\lambda)}{\phi(\lambda)}\frac{\varepsilon_{n+1}r_{n+1}^2-\varepsilon_n r_n^2}{r_{n+1}^2-r_n^2}-\frac{\eta_0}{2G_0}\frac{\eta(\lambda)}{\phi(\lambda)}\alpha\tilde{\lambda} \qquad (3.212)$$

The radial stresses on boundary surface $\sigma_r|_{r=r_n}=\sigma_r^+$ and $\sigma_r|_{r=r_{n+1}}=\sigma_{r+1}^+$ of the n-th binder with regard to swelling effect will equal:

$$\frac{1}{2G_0\phi(\lambda)}\sigma_n^+=\frac{\nu_0}{1-2\nu_0}\frac{\psi(\lambda)}{\phi(\lambda)}\varepsilon_z-\frac{1}{r_{n+1}^2-r_n^2}[r_{n+1}^2$$

$$+\left(1+\frac{2\nu_0}{1-2\nu_0}\frac{\psi(\lambda)}{\phi(\lambda)}\right)r_n^2]\varepsilon_n$$

$$+\frac{2r_{n+1}^2}{r_{n+1}^2-r_n^2}\left(1+\frac{\nu_0}{1-2\nu_0}\frac{\psi(\lambda)}{\phi(\lambda)}\right)\varepsilon_{n+1}-\frac{\eta_0}{2G_0}\frac{\eta(\lambda)}{\phi(\lambda)}\alpha\tilde{\lambda} \quad (3.213)$$

(for $n=1,2,...,N-1$)

$$\frac{1}{2G_0\phi(\lambda)}\sigma_{n+1}^{+}=\frac{\nu_0}{1-2\nu_0}\frac{\psi(\lambda)}{\phi(\lambda)}\varepsilon_z$$

$$-\frac{2r_n^2}{r_{n+1}^2-r_n^2}[1+\frac{\nu_0}{1-2\nu_0}\frac{\psi(\lambda)}{\phi(\lambda)}r_n^2]\varepsilon_n$$

$$+\frac{1}{r_{n+1}^2-r_n^2}[r_n^2+\left(1+\frac{2\nu_0}{1-2\nu_0}\frac{\psi(\lambda)}{\phi(\lambda)}\right)r_{n+1}^2]\varepsilon_{n+1}-\frac{\eta_0}{2G_0}\frac{\eta(\lambda)}{\phi(\lambda)}\alpha\tilde{\lambda} \quad (3.214)$$

(for $n=1,2,...,N-1$)

The axial stress interior to the layer is constant and equals (1.66). The adhesive strength of a reinforced cylinder with regard to swelling effect of the binding material is defined by the quantity of the jump of radial stress acting on the boundary of the n th load-bearing layer (braid, filament, and fabrics), since the jump of radial stress $\Delta\sigma_n$ is restrained by the force reinforcing layer. In this connection, it is established that the jump of radial stress on the n-th boundary surface with regard to swelling effect of the binding material is retrained by the tangential component of the force reinforcing layer $Y_n(\lambda)$ and equals:

$$\Delta\sigma_n=\sigma_n^{+}-\sigma_n^{-}=\frac{1}{r_n}Y_n(\lambda) \text{ (for } n=1,2,\ldots,N) \quad (3.215)$$

where $Y_n(\lambda)$ is the projection of tension forces of bundles (filaments) of the n-th layer on the axis y. Based on (3.214), define σ_n^-. For that in (3.215) change the index $(n+1)$ by n, and get:

$$\frac{1}{2G_0\phi(\lambda)}\sigma_n^- = \frac{\nu_0}{1-2\nu_0}\frac{\psi(\lambda)}{\phi(\lambda)}\varepsilon_z$$

$$-\frac{2r_{n-1}^2}{r_n^2 - r_{n-1}^2}[1+\frac{\nu_0}{1-2\nu_0}\frac{\psi(\lambda)}{\phi(\lambda)}]\varepsilon_{n-1}$$

$$+\frac{1}{r_n^2 - r_{n-1}^2}[r_{n-1}^2 + \left(1+\frac{2\nu_0}{1-2\nu_0}\frac{\psi(\lambda)}{\phi(\lambda)}\right)r_n^2]\varepsilon_n - \frac{\eta_0}{2G_0}\frac{\eta(\lambda)}{\phi(\lambda)}\alpha\tilde{\lambda} \quad (3.216)$$

(for $n = 2,3,\ldots,N$)

Equating (3.213) and (3.216) on the n-th boundary surface, we get:

$$\frac{1}{2G_0\phi(\lambda)}\Delta\sigma_n = \frac{1}{2G_0\phi(\lambda)}(\sigma_n^+ - \sigma_n^-)$$

$$= \frac{2r_{n-1}^2}{r_n^2 - r_{n-1}^2}[1+\frac{\nu_0}{1-2\nu_0}\frac{\psi(\lambda)}{\phi(\lambda)}]\varepsilon_{n-1}$$

$$-\{\frac{1}{r_{n+1}^2 - r_n^2}[r_{n+1}^2 + \left(1+\frac{2\nu_0}{1-2\nu_0}\frac{\psi(\lambda)}{\phi(\lambda)}\right)r_n^2]$$

$$+\frac{1}{r_n^2 - r_{n-1}^2}[r_{n-1}^2 + \left(1+\frac{2\nu_0}{1-2\nu_0}\frac{\psi(\lambda)}{\phi(\lambda)}\right)r_n^2]\}\cdot\varepsilon_n$$

$$+\frac{2r_{n+1}^2}{r_{n+1}^2 - r_n^2}\left(1+\frac{\nu_0}{1-2\nu_0}\frac{\psi(\lambda)}{\phi(\lambda)}\right)\cdot\varepsilon_{n+1} \quad (3.217)$$

(for $n = 2,3,...,N-1$)

Equalities (3.217) and (3.215) represents the contact equilibrium condition of the n th force reinforcing layer and ambient binder under different values of swelling parameter λ. The reinforcing load-bearing n th layer is thin and therefore in boundary conditions for a binder of thickness of a load-bearing layer we neglect it. Nevertheless, the n th load-bearing layer may consist of some ($i = 1, 2, \ldots$) sublayers with different angles β_{ni} to the generator. Define the density on the circle of the number of filaments of the i -th sublayer with the same angles by $2m_{ni}$. Then the general number of the filaments of the n -th load-bearing layer with different directions will equal (Figure 3.7 and 3.8):

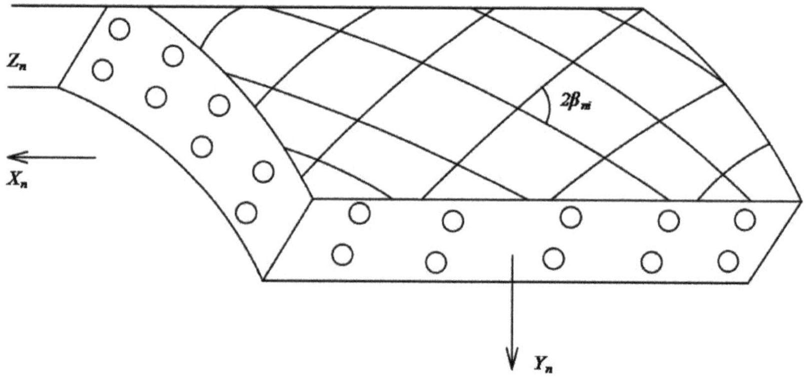

FIGURE 3.7 Scheme of load-bearing layer of reinforced pipe

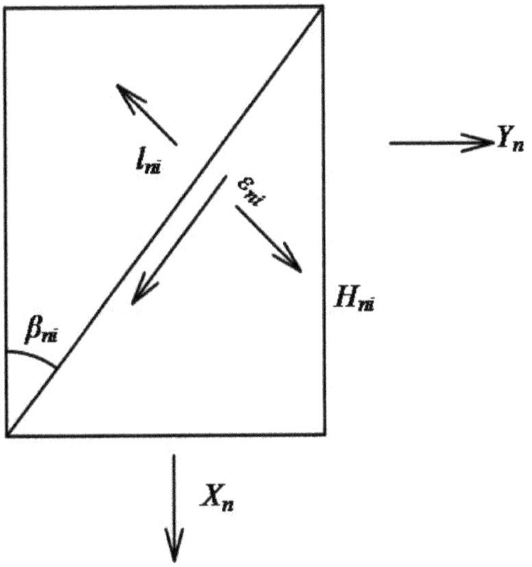

FIGURE 3.8 Relation of principal deformation on coordinate deformations

$$2\pi r_n \sum 2m_{ni} = (2m_{n1} + 2m_{n2} + ...)2\pi r_n \tag{3.218}$$

And the density of the filaments of the right $(\beta = +\beta_{ni})$ and the left $(\beta = -\beta_{ni})$ coilings are suggested to be the same and equal to m_{ni}. The step of the coilings of the n th layer equals $2\pi r_n \text{ctg}\beta_{ni} = H_{ni}$. Therefore, the density of the number of filaments along the generator with the coiling angles equals $\pm\beta_{ni}$ will equal $2m_{ni}\text{tg}\beta_{ni}$. Thus, $2m_{ni}$ and $2m_{ni}\text{tg}\beta_{ni}$ are the total numbers of filaments with their directions $\pm\beta_{ni}$ to the generator, per a unit length of the circle and the generator of the n th reinforcing layer. Then the total tensions of polymerized bundles (filaments, braids) along the axes x and y will equal:

$$X_n = 2\sum m_{ni} T_{ni} \cos\beta_{ni} \tag{3.219}$$

$$Y_n = 2\sum m_{ni} T_{ni} \text{tg}\beta_{ni} \cdot \sin\beta_{ni} \tag{3.220}$$

where T_{ni} is the tension of one polymerized braid with the angle β_{ni} $(or - \beta_{ni})$.

The bundle (filament) consisting of elementary fibers may be binded with a binder without slip. Then its longitudinal ε_ℓ and lateral $\varepsilon_\perp$ deformations will coincide with the deformations ε_ℓ^b and $\varepsilon_\perp^b$ of the binding material in the direction of the bundle (filaments) and perpendicular to it. In this case, the tension of the bundle (filament, braid) is determined both as a longitudinal deformation along the bundle and mean value of stress and deformation of lateral compression of the bundle, that is $T_H = \Phi(\varepsilon_\ell, \varepsilon_\perp, \sigma_\perp)$. For calculating the joint work of elementary fibers with a binder, we can make supposition on the following linear dependence of elastic tension of the bundle situated in the binding medium, on deformations with regard to change of physico–chemical properties of the binder:

$$T_f = (EF)_{ni}(\varepsilon_\ell + \nu_{\perp i}\varepsilon_{ni\perp} - \alpha_n\lambda) \tag{3.221}$$

Here ε_{ni} is the relative elongation of the binder in the direction β_{ni} of the bundle; $\varepsilon_{ni\perp}$ is the mean relative elongation of the binder in the lateral direction to the bundle $[E(\lambda)F(\lambda)]_{ni}$ is the tension rigidity of the bundle under the conditions of joint work with the binder with regard to swelling of the binding material; $\nu_{ni}(\lambda) \geq 0$ is the averaged coefficient of lateral compression of the bundle with regard to swelling parameter $[E(\lambda)F(\lambda)]_{ni}$ and $\nu_{ni}(\lambda)$ depend on impregnation density of the braid with the binder, adhesive bond forces and also on diffusion effect of corrosive liquid or gassy medium. They essentially depend on twisting degree of the bundle and at qualitative impregnation and adhesive bonds its rigidity diminishes according to increase of initial twisting of the bundle. Above we have considered experimental definition method of coefficients and verification of the suggested hypothesis (3.221) with regard to swelling effect.

For small deformations $\varepsilon_{ni} = \varepsilon\cos^2\beta_{ni} + \varepsilon_y\sin^2\beta_{ni}$, $\varepsilon_{ni\perp} = \varepsilon\sin^2\beta_{ni} + \varepsilon_y\cos^2\beta_{ni}$. Then the connection of the bundle's (filament's) stress T_{ni} with deformation of the n-th layer of the binder (3.221) will be of the form:

$$T_{ni}(\lambda) = E^{0}_{ni}(\lambda)F_{ni}(\lambda)\cdot\phi_{ni}(\lambda)[\varepsilon(\cos^2\beta_{ni} + v'_{ni}(\lambda)\sin^2\beta_{ni}) +$$
$$+\varepsilon_n(\sin^2\beta_{ni} + v_{ni}(\lambda)\cos^2\beta_{ni}) - \alpha_f\lambda] \tag{3.222}$$

Substituting (3.222) in (3.219) and (3.220), find the projections of total tensions of bundles (filaments) per a unit length of a circle and generator on the axes x and y with regard to the swelling parameter λ, in the form:

$$Z_n = B^z_{n1}\varepsilon + B^z_{n2}\varepsilon_n - B^z_{n0} \tag{3.223}$$

$$Y_n = B^y_{n1}\varepsilon + B^y_{n2}\varepsilon_n - B^y_{n0}. \tag{3.224}$$

Here,

$$B^z_{n1} = 2\sum m_{ni}E^0_{ni}F_{ni}(\lambda)\cdot\phi_{ni}(\lambda)(\cos^2\beta_{ni} + v_{ni}(\lambda)\sin^2\beta_{ni})\cos\beta_{ni}$$

$$B^z_{n1} = 2\sum m_{ni}E^0_{ni}F_{ni}(\lambda)\cdot\phi_{ni}(\lambda)(\sin^2\beta_{ni} + v_{ni}(\lambda)\cos^2\beta_{ni})\cos\beta_{ni}$$

$$B^y_{n1} = 2\sum m_{ni}E^0_{ni}F_{ni}(\lambda)\cdot\phi_{ni}(\lambda)(\cos^2\beta_{ni} + v_{ni}(\lambda)\sin^2\beta_{ni})\sin\beta_{ni}tq\beta_{ni}$$

$$B^y_{n2} = 2\sum m_{ni}E^0_{ni}F_{ni}(\lambda)\cdot\phi_{ni}(\lambda)(\sin^2\beta_{ni} + v_{ni}(\lambda)\cos^2\beta_{ni})\sin\beta_{ni}tq\beta_{ni}$$

$$B^z_{n0} = 2\alpha\lambda\sum m_{ni}E^0_{ni}F_{ni}(\lambda)\cdot\phi_{ni}(\lambda)\cdot\cos\beta_{ni}$$

$$B^y_{n0} = 2\alpha\lambda\sum m_{ni}E^0_{ni}F_{ni}(\lambda)\cdot\phi_{ni}(\lambda)\cdot\sin\beta_{ni}\cdot tq\beta_{ni} \tag{3.225}$$

Substituting (3.217) and (3.224) in (3.215), we get a contact condition on boundary surfaces $n = 2,3,\ldots,N-1$ with regard to change of physico–chemical properties of the binding material:

$$\frac{2r_{n-1}^2}{r_n^2 - r_{n-1}^2}[1+\frac{\nu_0}{1-2\nu_0}\frac{\psi(\lambda)}{\phi(\lambda)}]\varepsilon_{n-1} - \{\frac{1}{r_n^2 - r_{n-1}^2}[r_{n+1}^2$$

$$+\left(1+\frac{2\nu_0}{1-2\nu_0}\frac{\psi(\lambda)}{\phi(\lambda)}\right)r_n^2] + \frac{1}{r_n^2 - r_{n-1}^2}[r_{n-1}^2$$

$$+\left(1+\frac{2\nu_0}{1-2\nu_0}\frac{\psi(\lambda)}{\phi(\lambda)}\right)r_n^2]\}\cdot\varepsilon_n + \frac{2r_{n+1}^2}{r_{n+1}^2 - r_n^2}\left(1+\frac{\nu_0}{1-2\nu_0}\frac{\psi(\lambda)}{\phi(\lambda)}\right)\cdot\varepsilon_{n+1}$$

$$=\frac{1}{2G_0\phi(\lambda)\cdot r_n}[B_{n1}^{y}\varepsilon + B_{n2}^{y}\varepsilon_n - B_{n0}^{y}] \qquad (3.226)$$

The values of radial stresses and the first σ_1^- and the last σ_N^+ layers are determined from the boundary conditions:

$$r = r_1,\ \sigma_1^- = -p_a \qquad (3.227)$$

$$r = r_N,\ \sigma_N^+ = -p_b. \qquad (3.228)$$

Define the contact conditions on the first and the n th boundary surfaces with regard to change of physico–chemical properties of the binding material. For that we substitute (3.213), (3.227) and (3.244) in (3.215) for $n=1$, and also (3.216), (3.228) and (3.224) in (3.215) for $n=N$ and get contact conditions on the first and N th boundary surfaces with regard to change of physico–chemical properties of the binding material:

$$\frac{p_a}{2G_0\phi(\lambda)}+\frac{\nu_0}{1-2\nu_0}\varepsilon-\frac{1}{r_2^2-r_1^2}[r_2^2+\left(1+\frac{2\nu_0}{1-2\nu_0}\frac{\psi(\lambda)}{\phi(\lambda)}\right)r_1^2]\cdot\varepsilon_1$$

$$+\frac{2r_2^2}{r_2^2-r_1^2}\left(1+\frac{\nu_0}{1-2\nu_0}\frac{\psi(\lambda)}{\phi(\lambda)}\right)\cdot\varepsilon_2-\frac{\eta_0}{2G_0}\frac{\eta(\lambda)}{\phi(\lambda)}\alpha\tilde{\lambda}$$

$$=\frac{1}{2G_0\phi(\lambda)}\cdot\frac{1}{r_1}[B_{11}^y\varepsilon+B_{12}^y\varepsilon_1-B_{10}^y] \tag{3.229}$$

$$\frac{p_b}{2G_0\phi(\lambda)}+\frac{\nu_0}{1-2\nu_0}\frac{\psi(\lambda)}{\phi(\lambda)}\varepsilon-\frac{2r_{N-1}^2}{r_N^2-r_{N-1}^2}\left(1+\frac{\nu_0}{1-2\nu_0}\frac{\psi(\lambda)}{\phi(\lambda)}\right)\cdot\varepsilon_{N-1}$$

$$+\frac{1}{r_N^2-r_{N-1}^2}[r_{N-1}^2+\left(1+\frac{\nu_0}{1-2\nu_0}\frac{\psi(\lambda)}{\phi(\lambda)}r_N^2\right)]\cdot\varepsilon_N-\frac{\eta_0}{2G_0}\frac{\eta(\lambda)}{\phi(\lambda)}\alpha\tilde{\lambda}$$

$$=-\frac{1}{2G_0\phi(\lambda)}\cdot\frac{1}{r_N}[B_{N1}^y\varepsilon+B_{N2}^y\varepsilon_N-B_{N0}^y] \tag{3.230}$$

The longitudinal force P in the thin-walled sandwichly-reinforced pipe equals:

$$P=\pi\sum_{n=1}^{N-1}(r_{n+1}^2-r_n^2)\sigma_z+\sum_{n=1}^{N}2\pi r_n Z_n \tag{3.231}$$

Substituting (3.212) in (3.223), we get:

$$\frac{P}{2G_0\phi(\lambda)} = \left(1 + \frac{\nu_0}{1-2\nu_0}\frac{\psi(\lambda)}{\phi(\lambda)}\right)(r_N^2 - r_1^2)\cdot\varepsilon$$
$$+\frac{2\nu_0}{1-2\nu_0}\frac{\psi(\lambda)}{\phi(\lambda)}\cdot(\varepsilon_N r_N^2 - \varepsilon_1 r_1^2) - \frac{\eta_0}{2G_0}\frac{\eta(\lambda)}{\phi(\lambda)}\alpha\lambda\cdot(r_N^2 - r_1^2) + \tag{3.232}$$
$$+\frac{1}{G_0\phi(\lambda)}\cdot\sum_{n=1}^{N} r_n(B_{n1}^y\varepsilon + B_{nz}^y\varepsilon_n - B_{n0}^y)$$

The system (3.226), (3.229), (3.230) and (3.232) represents the $(N+1)$ equation with the same number of unknowns $\varepsilon_1, \varepsilon_2, \ldots, \varepsilon_N$ and ε.

3.7 STRESS–STRAIN STATE OF A THREE-LAYER REINFORCED PIPE UNDER THE ACTION OF INTERNAL PRESSURE AND AXIAL TENSION WITH REGARD TO PHYSICO–CHEMICAL CHANGE OF THE BINDING MATERIAL

Investigate a problem on the strength of a flexible pipe consisting of two layers of binding material and one reinforcing one located between them. The problem is solved under the assumption that physico–mechanical properties of the binding material depends on the change of its physico–chemical properties, that is from the swelling effect of the binding material arising when it contacts with corrosive liquid (Aliyev, 2012).

Define the quantities of stresses and deformations in reinforcing filaments, the contact stresses between the reinforcing and binding layers that define adhesive strength of the construction, and that arise from the joint action of external loads and the swelling degree of the binding material. The problem's solution is obtained from the system of equations (3.229), (3.230), (3.232) and (3.226), that conformity to the stated problem takes the form:

$$\frac{P}{2G_0\phi(\lambda)} + \frac{\nu_0}{1-2\nu_0}\varepsilon - \frac{1}{r_2^2 - r_1^2}[r_2^2 + \left(1 + \frac{2\nu_0}{1-2\nu_0}\frac{\psi(\lambda)}{\phi(\lambda)}\right)r_1^2]\cdot\varepsilon_1$$

$$+\frac{2r_2^2}{r_2^2-r_1^2}\left(1+\frac{\nu_0}{1-2\nu_0}\frac{\psi(\lambda)}{\phi(\lambda)}\right)\cdot\varepsilon_2-\frac{\eta_0}{2G_0}\frac{\eta(\lambda)}{\phi(\lambda)}\alpha\tilde{\lambda}$$

$$=\frac{1}{2G_0\phi(\lambda)}\cdot\frac{1}{r_1}[B_{11}^{y}\varepsilon+B_{12}^{y}\varepsilon_1-B_{10}^{y}] \tag{3.233}$$

$$\begin{aligned}&\frac{2r_1^2}{r_2^2-r_1^2}\left(1+\frac{\nu_0}{1-2\nu_0}\frac{\psi(\lambda)}{\phi(\lambda)}\right)\cdot\varepsilon_1\\&-\{\frac{1}{r_3^2-r_2^2}[r_3^2+\left(1+\frac{2\nu_0}{1-2\nu_0}\frac{\psi(\lambda)}{\phi(\lambda)}\right)r_2^2]+\frac{1}{r_2^2-r_1^2}[r_1^2\\&+\left(1+\frac{2\nu_0}{1-2\nu_0}\frac{\psi(\lambda)}{\phi(\lambda)}\right)r_2^2]\}\cdot\varepsilon_2+\frac{2r_3^2}{r_3^2-r_2^2}\left(1+\frac{\nu_0}{1-2\nu_0}\frac{\psi(\lambda)}{\phi(\lambda)}\right)\cdot\varepsilon_3\\&=\frac{1}{2G_0\phi(\lambda)}\cdot\frac{1}{r_2}[B_{21}^{y}\varepsilon+B_{22}^{y}\varepsilon_1-B_{20}^{y}]\end{aligned} \tag{3.234}$$

$$\begin{aligned}&\frac{\nu_0}{1-2\nu_0}\frac{\psi(\lambda)}{\phi(\lambda)}\cdot\varepsilon-\frac{2r_2^2}{r_3^2-r_2^2}\left(1+\frac{\nu_0}{1-2\nu_0}\frac{\psi(\lambda)}{\phi(\lambda)}\right)\cdot\varepsilon_2\\&+\frac{1}{r_3^2-r_2^2}[r_2^2+\left(1+\frac{2\nu_0}{1-2\nu_0}\frac{\psi(\lambda)}{\phi(\lambda)}\right)r_3^2]\cdot\varepsilon_3-\frac{\eta_0}{2G_0}\frac{\eta(\lambda)}{\phi(\lambda)}\alpha\tilde{\lambda}\\&=-\frac{1}{2G_0\phi(\lambda)}\cdot\frac{1}{r_3}[B_{31}^{y}\varepsilon+B_{32}^{y}\varepsilon_1-B_{30}^{y}]\end{aligned} \tag{3.235}$$

$$\begin{aligned}&[1+\frac{\nu_0}{1-2\nu_0}\frac{\psi(\lambda)}{\phi(\lambda)}](r_3^2-r_1^2)\cdot\varepsilon+\frac{2\nu_0}{1-2\nu_0}\frac{\psi(\lambda)}{\phi(\lambda)}(\varepsilon_3 r_3^2-\varepsilon_1 r_1^2)\\&-\frac{\eta_0}{2G_0}\frac{\eta(\lambda)}{\phi(\lambda)}\alpha\lambda\cdot(r_3^2-r_1^2)\\&+\frac{1}{G_0\phi(\lambda)}\cdot\sum_{n=1}^{3}r_n\cdot[B_{n1}^{y}\varepsilon+B_{n2}^{y}\varepsilon_1-B_{n0}^{y}]=\frac{P}{2\pi G_0\phi(\lambda)}\end{aligned} \tag{3.236}$$

The system of equations (3.233) and (3.236) is a linear-algebraic system of four equations for the deformations $\varepsilon_1, \varepsilon_2, \varepsilon_3, \varepsilon_z = \varepsilon_4$. Write this system in the matrix form:

$$a_{ij}\varepsilon_j = b_i \tag{3.237}$$

Then by the Kramer's rule, the solution of the linear algebraic system (3.237) will equal:

$$\varepsilon_j = \frac{\left|D_{\varepsilon_j}\right|_4}{\left|a_{ij}\right|_4} = \frac{\sum_{i=1}^{4}(-1)^{i+j}\cdot b_i \cdot B_{ij}^{\varepsilon_j}}{\sum_{k=1}^{4}(-1)^{1+k}\cdot a_{1k}\cdot A_{1k}} \tag{3.238}$$

In the expanded form (3.238) will take the form:

$$\varepsilon_1 = \frac{\left|D_{\varepsilon_j}\right|_4}{\left|a_{ij}\right|_4} = \frac{\sum_{i=1}^{4}(-1)^{i+1}\cdot b_i \cdot B_{i1}^{\varepsilon_1}}{\sum_{k=1}^{4}(-1)^{1+k}\cdot a_{1k}\cdot A_{1k}},$$

$$\varepsilon_2 = \frac{\left|D_{\varepsilon_j}\right|_4}{\left|a_{ij}\right|_4} = \frac{\sum_{i=1}^{4}(-1)^{i+2}\cdot b_i \cdot B_{i2}^{\varepsilon_2}}{\sum_{k=1}^{4}(-1)^{1+k}\cdot a_{1k}\cdot A_{1k}},$$

$$\varepsilon_3 = \frac{\left|D_{\varepsilon_j}\right|_4}{\left|a_{ij}\right|_4} = \frac{\sum_{i=1}^{4}(-1)^{i+3}\cdot b_i \cdot B_{i3}^{\varepsilon_3}}{\sum_{k=1}^{4}(-1)^{1+k}\cdot a_{1k}\cdot A_{1k}},$$

$$\varepsilon_4 = \frac{\left|D_{\varepsilon_j}\right|_4}{\left|a_{ij}\right|_4} = \frac{\sum_{i=1}^{4}(-1)^{i+4}\cdot b_i \cdot B_{i4}^{\varepsilon_4}}{\sum_{k=1}^{4}(-1)^{1+k}\cdot a_{1k}\cdot A_{1k}}$$

Having given the mechanical and physico–chemical properties of the binder and reinforcing layer and also geometrical characteristics of the reinforced construction r_i, β_i, the boundary deformations become uniquely defined. When there is no swelling effect, the solution of the problem co-incides, with the solution obtained in (Aliyev, 1987).

KEYWORDS

- **Elastic constants**
- **Polymer matrix**
- **Quasi-static interaction**
- **Reinforced pipe**
- **Sandwich pipe**

REFERENCES

1. Akhenbakh, J. D. (1965). The dynamic behavior of a long fastened to the body of visco-elastic cylinder. Raketnaya Tekhnika I kosmonaftika 4.
2. Aliyev, G. G. (1987). Fundamentals of mechanics of reinforced flexible pipes (p. 265). Baku: Elm.
3. Aliyev, G. G. (1984). On the physical and mechanical properties of the beam shaped bodies with quasi-static forces with the matrix (Issue 5). Baku: Elm.
4. Aliyev, G. G. & Gabibov, I. A. (1994). Resistance of polymeric and composite materials for the changes in the physico-chemical properties (p. 167). Baku: Azeri. Oil Academy.
5. Aliyev, G. G. (1995). Polimer və Kompozit Malzemelerin Dirençlerinin Fiziko-Kimyasal Özəlliklərİyle Değişimi Teorisi. Milli Eğitim Bakanlığı Yayınları 2963, "Bilim və Kültür Eserleri Dizisi: 871". Eğitim Dizisi-№5, Yayın Kodu 95.34.Y.0002.1376, İSBN 975.11.1046-7, Türkiyə, İstanbul, 119 sayfa.
6. Aliyev, G. G. (1998). Kompozit maddeler mekaniginin matematik temelleri. NÜ yayınları, No. 2, Türkiyə, Istanbul, ISBN: 975-8062-03-4, p. 563.
7. Aliyev, G. G. (2009). Longitudinal vibration of polymerized bar with regard to lateral motion dynamics. Azerbaijan National Academy of Sciences, Transaction issue mathematics and mechanics series of physical-technical and mathematical science (No. 4, p. 179–186). Baku: Elm.
8. Aliyev, G. G. (1980). Elastic-plastic deformation of flexible multi-layered, reinforced thick-walled cylinders (p. 80). Baku: Space Science.
9. Aliyev, G. G. (2011). On a mechanical model of deformation polymerized bundle of fibers in the cervix. The International Scientific Symposium on mechanics of deformable bodies, dedicated to the 100th anniversary of the birth of A. A. Ilyushin, pp. 292–295, Moscow.
10. Aliyev, G. G. (2012). Eexperimental-theoretical method for defining physical-mechanical properties of polyimer materials with regard to change of their physical-chemical properties. usa, Iinternational Journal of Chemoinformatics and Chemical Engineering (IJCCE), 2(1), 12–24, January–June.
11. Aliyev, G. G. (2012). Pphysical-mechanical model of fibers polimerized bar deformation at the neck zone. USA, International Journal of Chemoinformatics and Chemical Engineering (IJCCE), 2(2), 12–24, July–December.

12. Aliyev, G. G. (2012). Fundamentals of mechanics of polymeric and composite materials with regard to changes in the physico-chemical properties. Ed. "Elm" ANAS, pp. 286., ISBN 978-9952-453-13-3.
13. Barrer, M., Jaumotte, A., Vebek, B. F., & Vanderkerhove, J. (1962). Rocket engines. Translated from English. Moscow: Oborongiz.
14. Durelli, A. J. (1967). Experimental strain and stress analysis of solid propellant rocket motors, Mechanical and chemical solid propellants., Oxford: Oxford University Press.
15. Grot, S. & Mazur, P. (1967). Nonequilibrium thermodynamics (p. 456). Moscow: Mir.
16. Eroshenko, V. M., Zaychik, L. I., & Zorin, I. B. (1980). Calculation of mechanical resistance and heat and mass transfer in laminar flow of an incompressible fluid with variable physical properties in pipes with porous walls. Engineer-fis. Magazine, 39(3), 462–467.
17. Eroshenko, V. M. & Zaychik, L. I. (1984). Hydrodynamics and heat transfer on permeable surfaces (p. 273). Moscow: Nauka.
18. Jukov, A. M. (1949). On the neck of the sample in tension. Engineer Collection, 5(2).
19. Ferry, J. (1963). Viscoelastic properties of polymers. Moscow: IL.
20. Ilyushin A. A. (1940). Proceedings of the Moscow State University named M.V. Lomonosov, 39, (in Russian), Moscow (in Russian).
21. Ilyushin, A. A. (2011). Infinitely close motions and stability of the deformation of visco-plastic body. Tension-compression band and close movement. Proceedings of the International Symposium on MRTT dedicated to the 100th anniversary of the Ilyushin A.A., Ed. Moscow State University,. MSU, pp. 147-–195, (in Russian) Moscow (in Russian).
22. Ilyushin, A. A. (1948). Plasticity. Part 1. Elastic-plastic deformations (in Russian). Moscow: Gostekhizdat (in Russian).
23. Ilyushin, A. A. & Lenskiy, V. S. (1959). Strength of materials. (in Russian) Moscow: Fizmatgiz. (in Russian).
24. Ilyushin, A. A. & Pobedrya, B. E. (1970). Foundations of mathematical theory of thermoviscoelasticity (in Russian p. 290). Moscow: Nauka (in Russian).
25. Ferry J. (1963). Viscoelastic Properties of Polymers (in Russian). Moscow: IL.
26. Ishlinskiy, A. Yu. (1943). Stability of viscoplastic flow strip and circular rod (in Russian Vol. 7, No. 2). Moscow: PMM (in Russian).
27. Kargin, V. A. & Slonimsky, G. L. (1960). Brief essays on the physical chemistry of polymers Publishing House of Moscow State University (in Russian). Moscow: Lomonoso Moscow State University (in Russian).
28. Korten, B. A. (1967). Destruction reinforced plastics. Trans. Translated from English. Moscow: IL Chemistry.
29. Lykov, A. V. (1978). Heat and mass transfer (p. 479). Moscow: Energiya.
30. Manin, V. A. & Gromov, A. (1980). Physical and chemical stability of polymeric materials in the field (p. 248). Moscow: IL Chemistry.
31. Malmeyster, A. K., Tamuzh, V. P., & Teters, G. A. (1980). Resistance of polymeric and composite materials. Riga: Zinaite.
32. Mac-Gregor, C. W. (1944). Tthe true streess-strain tension test its rolle in modern materials testing. Journal of the Franklin Institute, 2–3, 238.
33. Moskvitin, V. V. (1972). Strength of viscoelastic materials (p. 327)(in Russian Fiz.-mat. giz p. 327). Moscow: Nauka (in Russian Fiz.-mat. giz).

34. Ogibalov, P. M. & Suvorov, Y. V. (1965). Mechanics of reinforced plastics (in Russian p. 480). Moscow: MGU (in Russian).
35. Ogibalov, P. M., Malinin, N. I., Netrebko, V. P., & Kishkin B. P. (1972). Engineering polymers (in Russian Vol. 2, p. 306). Moscow: MGU (in Russian).
36. Reitlinger S.A. (1970). Permeability of polymer materials (p. 479). Moscow: Chemistry.
37. Sachs, G. & Lubahn, J. (1946). Ffailure of ductile metals in tension. Transactions ASME, 4, 68.
38. Shapiro, J. M., Mazing, G. Y., & Prudnikov, N. E. (1968). Principles of design for solid fuel rockets (in Russian). Moscow: Military Publishers (in Russian).
39. Shen, F. C., Chen, T. S., & Huang, L. M. (1976). The effects of main-flow radial velocity on the stability of developing laminar pipe flow. Journal of Applied Mechanics, 43(2), 209–213.
40. Siebel, E. (1944). Zur Mechanik des Zugversuchs (p. 2). Folge: Werkstofforschung.
41. Sedov, L. I. (1970). Continuum mechanics (in Russian Vol. 1, 2). Nauka: Moscow (in Russian).
42. Sorokin, R. E. (1967). Gas dynamics of solid propellant rocket engine (in Russian). Nauka: Moscow (in Russian).
43. Starovoitov, E. I. & Nagiyev, F. B. (2012). Foundations of the theory of elasticity, plasticity and viscoelasticity (p. 346). Toronto: Apple Academic Press.
44. Stepanov, R. D. & Shlenscy, O. F. (1981). Strength calculation of structures made of plastic, operating in liquid media (p. 370). Moscow: Engineering.
45. Summerfield, M. (Eeds.). (1963). Investigation of solid propellant rocket engine. Moscow: IL.
46. Tikhomirov, N. S. (1970). Aggressive factors influence the diffusion of substances. Moscow: Chemistry.
47. Tynny, A. N. (1975). Strength and fracture of polymers under the influence of liquid corrosion (p. 206). Moscow: Naukova Dumka.
48. Urzhumtseva, Y. S. (1982). Prediction of long-term strength of polymeric materials. Moscow: Nauka.
49. Zuev, B. S. (1972). Polymer degradation under the action of aggressive media (p. 235). Moscow: Khimiya.

INDEX

Milton Keynes UK
Ingram Content Group UK Ltd.
UKHW050258161024
449569UK00042B/1786

9 781774 632840